PICTURE INTERPRETATION
A Symbolic Approach

SERIES IN MACHINE PERCEPTION AND ARTIFICIAL INTELLIGENCE*

Editors: **H. Bunke** (Univ. Bern, Switzerland)
P. S. P. Wang (Northeastern Univ., USA)

Vol. 8: Thinning Methodologies for Pattern Recognition
(Eds. *C. Y. Suen and P. S. P. Wang*)

Vol. 9: State of the Art in Digital Mammographic Image Analysis
(Eds. *K. W. Bowyer and S. Astley*)

Vol. 11: Experimental Environments for Computer Vision and Image Processing
(Eds. *H. I. Christensen and J. L. Crowley*)

Vol. 12: Computing Conceptual Organization in Computer Vision
(Eds. *S. Sarkar and K. L. Boyer*)

Vol. 13: Progress in Automatic Signature Verification
(Eds. *R. Plamondon*)

Vol. 14: Document Analysis Systems
(Eds. *A. Dengel and L. Spitz*)

Vol. 15: Parallel Image Analysis and Processing
(Eds. *K. Inoue, A. Nakamura, M. Nivat, A. Saoudi and P. S. P. Wang*)

Vol. 16: Document Image Analysis
(Eds. *H. Bunke, P. S. P. Wang and H. Baird*)

Vol. 17: Applications of AI, Machine Vision and Robotics
(Eds. *K. L. Boyer, L. Stark and H. Bunke*)

Forthcoming

Vol. 10: Generic Object Recognition Using Form and Function
(Eds. *K. W. Bowyer and L. Stark*)

Vol. 18: VLSI and Parallel Computing for Pattern Recognition and AI
(Ed. *N. Ranganathan*)

Vol. 19: Parallel Image Analysis: Theory and Applications
(Eds. *L. S. Davis, K. Inoue, M. Nivat, A. Rosenfeld and P. S. P. Wang*)

Vol. 20: Picture Interpretation: A Symbolic Approach
(Eds. *S. Dance, T. Caelli and Z.-Q. Lin*)

Vol. 21: Modelling and Planning for Sensor Based Intelligent Robot Systems
(Eds. *H. Bunke, T. Kanade and H. Noltemeier*)

*For the complete list of titles in this series, please write to the Publisher.

Series in Machine Perception and Artificial Intelligence – Vol. 20

PICTURE INTERPRETATION
A Symbolic Approach

Sandy Dance
Department of Computer Science
University of Melbourne

Terry Caelli
Department of Computer Science
Curtin University of Technology

Zhi-Qiang Liu
Department of Computer Science
University of Melbourne

World Scientific
Singapore • New Jersey • London • Hong Kong

Published by

World Scientific Publishing Co. Pte. Ltd.

P O Box 128, Farrer Road, Singapore 9128

USA office: Suite 1B, 1060 Main Street, River Edge, NJ 07661

UK office: 57 Shelton Street, Covent Garden, London WC2H 9HE

PICTURE INTERPRETATION: A SYMBOLIC APPROACH

ISBN 981-02-2402-8

This book is printed on acid-free paper.

Printed in Singapore by Uto-Print

Preface

Traditional methods for image scene interpretation and understanding are based mainly on such single-threaded procedural paradigms as hypothesize-and-test or syntactic parsing. As a result, these systems are unable to carry out tasks that require concurrent hypothesis testing.

In this book we explore a method for symbolically interpreting images based upon a parallel implementation of a network-of-frames suggested, for example, by Minsky (1975), to describe intelligent processing. The system has been implemented in an object-oriented environment in the logic programming language Parlog++ and includes the propagation of uncertainty through each frame (object) using Baldwin's (1986) formulation. The system is tested with several scenarios of increasing complexity, culminating with the legal interpretation of traffic intersection images.

The authors would like to thank Andrew Davison for much help with the implementation language of this project, and for providing exceptional advice on ideas, papers and this book. We must also acknowledge the Computer Vision and Machine Intelligence Laboratory at Melbourne University as a whole for providing a stimulating and lighthearted atmosphere for intellectual jousting. The Computer Science Department must be recognized for providing a supportive role in the work involved. We would like to acknowledge the Australian Computer Graphics Centre of the Royal Melbourne Institute of Technology, and in particular Gordon Lescinsky and Mike Gigante, for their help in obtaining traffic video and digitizing it. Finally, the first author would like to thank his wife Jill and children Felix and Rupert for their tolerance of this folly.

This work has been supported by a number of funding sources and employers. The Defense Science and Technology Organization supported the early stages of this work. Funding has been received from the Centre for Intelligent Decision Systems, and from the Computer Science Departments at the University of Melbourne and Curtin University. Finally we would like to acknowledge the support of the James Herbert Miriam Bursary through the Faculty of Engineering at the University of Melbourne.

S.D.

T.M.C.

Z.Q.L.

Melbourne, April, 1995.

Contents

1 Introduction   **1**

2 A Recent History of Image Interpretation   **4**

    2.1 Graph Based Systems 4

         2.1.1 Augmented Transition Networks 5

         2.1.2 Semantic Networks 6

         2.1.3 Constraint Propagation Networks 6

         2.1.4 Discussion 7

    2.2 A German School 8

    2.3 Object Oriented Approaches 9

    2.4 Other Recent Systems 10

3 Foundations   **14**

    3.1 Human Conceptualization 14

    3.2 Frames and Related Ideas 16

    3.3 Logic Programming 20

    3.4 Object Orientation 21

4 Architecture of SOO-PIN   **24**

    4.1 Procedural Description of SOO-PIN 26

    4.2 Spatial Data 29

    4.3 Compound Objects 30

    4.4 Concept-Frame Structure 31

5 **Simple Scenarios** **35**
  5.1 Cutlery Scenario 35
  5.2 Trial Runs of Cutlery Scenario 38
  5.3 Wheels Scenario 42
  5.4 Trial Runs of Wheels Scenario 44
  5.5 Summary 47

6 **Interpretation of Traffic Scenes** **48**
  6.1 Primitive Concept-Frames 51
  6.2 Turn Concept-Frames 51
  6.3 Give-Way Concept-Frames 53
  6.4 Trial Runs using XFIG 53
  6.5 Low-Level Processing 60
  6.6 Trial Runs on Real Images 63
  6.7 Traffic Trial Summary 64

7 **Uncertainty** **66**
  7.1 Introduction 66
  7.2 Dempster-Shafer Theory 68
    7.2.1 Combining Evidence within a Frame of Discernment . 70
    7.2.2 Combining Independent Propositions 71
      7.2.2.1 Conjunction Rule 72
      7.2.2.2 Disjunction Rule 74
      7.2.2.3 Plausibility 76
      7.2.2.4 Combining Belief from Two of N Events . . . 77
  7.3 Problems with Dempster-Shafer 80
  7.4 SOO-PIN and Uncertainty 81
    7.4.1 Belief and Vision 81
    7.4.2 Implementation of Uncertainty in SOO-PIN 82
      7.4.2.1 Data Structures 82
      7.4.2.2 Existence Checking 82
      7.4.2.3 Belief Updating 83

	7.4.2.4	Procedural Subroutines	84
	7.4.2.5	Belief Runtime Experiments	85
7.5	Summary		87

8 Velocity   **88**

8.1	Introduction	88
8.2	Finding Trajectories	90
	8.2.1 Matching between Frames	91
	8.2.2 Finding Trajectories by Comparing Match Lists	92
	8.2.3 Determining Velocity	93
8.3	Uses of Velocity	94
8.4	Velocity Examples	98
8.5	Summary	102

9 Runtime Results   **103**

9.1	Introduction	103
9.2	Trial Runs	103
9.3	Summary of Results	106

10 Conclusion   **119**

Appendix A

Parlog++ Procedures	122
A.1 Switchboard Source Code	122
A.2 Give-Way Source Code	126

References   **132**

Index   **140**

List of Figures

3.1 The communication channel in Parlog 22

4.1 The SOO-PIN system concept 27
4.2 The implemented system architecture 28

5.1 The cutlery scenario network-of-frames 36
5.2 The Wheels Scenario network 43

6.1 Typical traffic scene processed by the SOO-PIN system 49
6.2 The traffic scenario network-of-frames 50
6.3 Diagram showing how turn.c determines car activity 52
6.4 Diagram showing a schematic intersection 54
6.5 Diagram showing a schematic intersection with lights 56
6.6 Diagram showing a schematic T-intersection 57
6.7 Diagram showing a section of road 59
6.8 Processing steps of real image 62
6.9 Real image and labeled image 64

7.1 Row of cars straddling a boundary demonstrating drop in belief 86

8.1 Example of 3 successive frames with the movement of one car 90
8.2 Algorithm for finding matches between cars in successive frames 91
8.3 Velocity vectors between a pair of cars in successive frames . . 92
8.4 Diagram showing a potential bad match 93
8.5 Diagram showing 3 cars turning right, over 3 frames 94

8.6 The expanded network with the concept-frames dealing with velocity 95

8.7 The geometry involved in calculating collisions 96

8.8 Example of velocity from 4 cars in 3 frames 99

8.9 Example of velocity from 4 cars in 3 frames in curving trajectory 100

8.10 Example of detection of a collision between two cars 101

9.1 Cars in T-intersection, XFIG diagram 108

9.2 Cars in T-intersection, XFIG diagram 109

9.3 Collins & Exhibition Sts., frame 89 110

9.4 Collins & Exhibition Sts., frame 116 111

9.5 Lygon & Queensberry Sts., frame 82 112

9.6 Lygon & Queensberry Sts., frame 130 113

9.7 Lygon & Queensberry Sts., frame 150 114

9.8 Lygon & Queensberry Sts., frame 190 115

9.9 Swanston & Faraday Sts., frame 120 116

9.10 Swanston & Faraday Sts., frame 242 117

9.11 Swanston & Faraday Sts., frame 316 118

List of Tables

2.1 Survey of status of vision systems, pre 1990 12

2.2 Survey of status of vision systems, post 1990 13

4.1 Table showing the actions performed and the messages sent by a concept-frame 34

7.1 Evidence combinations for the burglar example 71

7.2 Masses assigned to the members of $\Theta \times \Phi$ with non-zero mass 75

7.3 Example of the algorithm belTwoOfN in action 79

Chapter 1

Introduction

This book concerns high-level vision – the interpretation of images in the light of domain knowledge. In this realm, it is not sufficient to simply label regions or objects in the image. Rather, it is necessary to "tell the story" **behind** the image, ie, to interpret the "intentionality" of the image objects. This book explores a specific architecture for this problem: an interacting network of agents or frames, each asynchronously working on their own aspect of the interpretation. The architecture has been named SOO-PIN: symbolic object-oriented picture interpretation network. It is based upon recent developments in:

- Cognitive science,

- Object-oriented software engineering,

- Connectionism in Artificial Intelligence.

The approach taken in this study has been to investigate the architecture in a series of domains of increasing complexity. Initially a very simple domain was chosen in order to design the basic modules and communication protocols. Once that was successful, the system was adapted and expanded to determine scaling of the concepts involved. Success was measured in both correctness of interpretation for a given scene, and ease of adaptability of the system to

1

changing domain knowledge. These studies culminated in the analysis of a complex traffic scene analysis system.

In the course of the study a number of specific ideas were developed:

- The use of an object-oriented concurrent logic programming language for scene analysis [19].

- The use of the above architecture for traffic scene analysis [18].

- The modification of Baldwin's support logic programming for use in scene analysis [20].

- Distinguishing symbolic and spatial data, and recognizing the need to ground some symbolic predicates in the data. This reflects, in some ways, the distinction between internal and external representations as discussed by Slezak [63].

- For the interpretation of traffic, a velocity detection mechanism was developed which is founded on token matching but using a combination of symbolic and spatial predicate matching to derive the best trajectory for the vehicles [20].

Chapter 2 is a description of a number of earlier high-level vision systems which are essentially single-threaded approaches using either some form of syntactic parsing (ie, using transition networks) or hypothesize and test procedures. Multi-threaded systems are then discussed ending with some similar approaches to that described here, together with their drawbacks – which have motivated the approach adopted in this book.

Chapter 3 deals with the foundations of the approach taken here, drawing on the work of Herskovits, Lakoff, Minsky, Schank and Hewitt. The chapter ends with a review of logic programming and object-orientation, describing the work of Shapiro and Takeuchi which unites these two paradigms, and culminating in the choice of an implementation language, Parlog++.

Chapter 4 describes in detail the investigation and implementation of the system, its data structures, how the modules (here called "concept-frames" to distinguish them from frames, classes and objects as used in the literature) were implemented in Parlog++, and how they were networked together into a message passing system.

Chapter 5 describes the first two simple domains used for testing the system, the "cutlery" and the "wheels" scenario. **Chapter 6** then goes onto describe the initial "traffic" scenario, which uses video images of intersections as input. The hierarchy of "concept-frames" used to deal with this domain is described, together with the necessary low-level processing.

Chapter 7 explores the incorporation of "belief" or "uncertainty" into the network – from a vision perspective. Specifically, Baldwin's approach has been adopted (with reservations and extensions). The chapter finishes with a few trial runs of the system utilizing the derived uncertainty measure, together with analysis and summary.

Chapter 8 describes how the traffic scenario was enhanced with the calculation of car velocities. It first reviews some of the literature on velocity determination and the correspondence problem. It goes on to describe the technique used in SOO-PIN using the "best match between 3 frames" technique, and its implementation.

Chapter 9 is the culmination of the study, containing a series of analyzed traffic scenes, each one discussed at length, highlighting the systems successes and failures.

Chapter 10, the conclusion, points out the generality of the system, and its limitations, together with desirable extensions – the directions research should continue in the future.

Chapter 2

A Recent History of Image Interpretation

Computational vision has traditionally focused on problems of low-level sensing, pattern and object recognition, and relatively little attention has been paid to making sense of an image once primitive structures have been identified. This entails fitting the image into a broader context, or equivalently, bringing to bear domain knowledge in order to produce an interpretation. Such domain knowledge is typically symbolic and so the problem involves interpreting numerical data with more logical or syntactic information. The problem, then, reduces to that of finding the best architecture for integrating this domain knowledge with the labeled image data. Previous approaches have been based on the single-threaded graph processing programs, using hypothesize-and-test or syntactic parsing. Some of these "picture language" systems reported in the literature are described below. More recently there has been a move to implement systems based on the object-oriented paradigm, and these are described later in this chapter.

2.1   Graph Based Systems

These systems use some form of graph parsing or traversing algorithm to process the image. This can take place in top-down or bottom-up directions.

They can also be characterized as single-threaded or multi-threaded. Single-threaded systems, as used here, are those in which at any one moment there is at most a single hypothesis under consideration, whereas multi-threaded systems entertain more than one hypothesis simultaneously. This single- vs. multi-threaded distinction is not the same as the sequential vs. concurrent architecture distinction introduced later.

2.1.1 Augmented Transition Networks

The first technique considered is the augmented transition network (ATN) [14, p199]. This technique was developed for parsing natural language, but has been adapted for vision. It is based on the notion of states (the graph nodes) and the paths for moving between states (the arcs between nodes). The traversal through the graph is determined by the sequence of tokens (language elements or image segments) on the input. It is a top-down technique because it starts with the graph describing a full sentence or pattern, and invokes subgraphs to parse nonterminal symbols as required. The algorithm is augmented with stored data from past states which is used to influence decisions.

Tropf and Walters [64] describe a vision system which is one of the earliest to employ an ATN. It is used to control an analysis-by-synthesis approach (hypothesis-and-test), with the augmentations storing the model-to-pattern associations, ie, the 3D space constraints upon the object. The ATN describes the structure of 3D objects, which the system locates in the input image.

Bajcsy et al[5] describe a system that takes 3D polyhedral shapes derived from aerial photographs of urban scenes and uses an ATN to perform high level scene analysis, guided by user queries. This system approaches the vision problem as analogous to language, and treats the world knowledge as a grammar, and the segmented faces and the relations between them as a dictionary.

2.1.2   Semantic Networks

The semantic network (or associative network) [14] is a way of representing a collection of concepts (nodes) and the relationships between them (the arcs). In fact, the predicate calculus is an equivalent representation, but procedurally, the semantic network differs from predicate calculus in that it indexes the relationships by their endnodes. This optimizes traversal of the network from concepts through relationships to other concepts.

Nieman et al[51] describe a system that uses a semantic net to model relations between concepts in image sequences of the human heart, and use production rules to draw inferences (diagnostic descriptions). The system creates a "search tree" of competing instances (multiple hypotheses) at each node of the network, and uses an A* search for overall control.

Another medical system is that of Dellepiane et al[23] who use a semantic network to model the relations between parts of 2D and 3D models of brain tomographic imagery. They use the hypothesize-and-test paradigm constrained by network relations, and common data is stored in a blackboard. The output is labeled regions.

Govindaraju et al[33] use a top down approach based on a semantic network to handle one aspect of their multi-modal system for understanding newspaper images. This aspect relates 2D newspaper photographs to their captions, and, for instance, finds named people in the image. They also use an augmented transition network (ATN) to parse caption text, generating part of the semantic network used in processing the image. The rest of the network derives from domain knowledge.

2.1.3   Constraint propagation networks

These are networks in which nodes represent propositions, and arcs represent dependencies, or justification links, between them. Constraint propagation networks are logically equivalent to the propositional calculus (as opposed to predicate calculus) as there is no way to represent quantified expressions [68].

Mulder et al[49] describe a series of refinements of the MAPSEE system that, given a 2D geographic sketch map constrained by syntactic rules and represented by plotter commands, uses local constraint propagation over a graph to handle labeling hypotheses. Reiter and Mackworth[55] use the MAPSEE system as a basis for analyzing vision problems in terms of logical frameworks. From this perspective, vision problems reduce to theorem proving on a first-order logic database consisting of prior knowledge, image domain knowledge, constraints derived from contingent knowledge about the scene, and, finally, queries about the scene. One limitation of this perspective is that many spatial predicates can not be computed symbolically, and so the interpretation problem can not be exclusively cast as a logic program. This is further discussed in Section 4.2.

Boddington et al[10] also use constraint propagation, in this case an assumption-based truth maintenance system (ATMS), to maintain partial results of competing interpretations. They develop a system, CARRS, and show that the system is able to locate cars in some natural scenes. Provan, however, has shown that such ATMSs generally have exponential complexity [54]. Nevertheless the system is feasible in this case because of judicious early pruning of hypotheses.

2.1.4 Discussion

Representing problems in terms of graphs, while logically equivalent to predicate calculus, is a good way of representing problems for humans, in so far as the logical links between parts of the problem become clear. Moreover, the graphical representation suggests ways of solving problems which can, in some cases, be more efficient than logic-based approaches (these are discussed later in Section 3.3).

All these systems (with the exception of that of Mulder et al [49] and Nieman et al [51]) are however, single-threaded (here meaning that there is at most one hypothesis under consideration at any one time) which reduces the range of problems they can handle (or alternatively, increases the complexity). Moreover, the systems are complex, non-modular constructions which impede easy understanding and implementation.

2.2   A German School

In Germany a number of related projects have been exploring high-level vision based on sequences of images, and outputting German language interpretations.

One example is the traffic interpretation system of Schirra et al [61]. The scene is segmented by motion vectors resulting in identified objects in the sequence, which are described in terms of enclosing rectangles corners and displacement vectors. The output from the low level system is described in terms of static and dynamic objects (geometric scene description or GSD). This data is fed to the language system CITYTOUR ([56]), in which the user is assumed to be in the scene, and interacts with the language system which answers questions about the scene from the users viewpoint. This system handles static spatial relationships, for example, *in front of, behind, to the right* and the procedures then produce degrees of applicability (belief) for the relationships which are used for linguistic hedges (words like *maybe, possibly*). It should be noted that the static data is hand generated, and only the dynamic data is computer generated. Further, the link between the vision and language subsystems is one way (no feedback) and are, in fact, connected via TCP/IP between different cities !

Another approach similar to Schirra et al's is SOCCER (Andre et al [4]) which generates soccer descriptions of image sequences from soccer games. Static parts of the scenes are manually generated, dynamic parts are fed in from a hypothetical vision system, in the form of a geometric scene description (same terminology as Schirra et al's). They employ a data driven bottom-up strategy, using case models to recognize events using a transition net, which, in turn, triggers language output.

Neumann [50], takes 3D geometric scene description sequences, which contain data on the time, location and orientation of objects found in traffic scenes (but actually hand generated), and use a failure-driven / backtracking algorithm on a semantic network to generate case frames (or thematic roles, an objects thematic role specifies the objects relation to an act [68]). These frames are used as the interface to a more formal logic programming language for the high level interpretations.

These German systems are grouped together because they use the same representation for the low-level data (GSD), and generate a similar level of output (German language). They all use a top-down strategy for generating the language output, in the first case through interacting with user queries, and in Neumann's case through his semantic net and case frame techniques.

However, like the graph-based systems discussed above, these systems also suffer from complexity, containing a number of sophisticated subsystems using a variety of different techniques working together via heuristic constraints. One way of overcoming such cases of complexity (and difficulties with software) is to develop systems with intrinsic concurrency. The object-oriented approaches sets up the framework for such systems.

2.3 Object Oriented Approaches

In the 1990s, a new generation of vision systems was developed, those based on the object-oriented paradigm. This approach overcomes some problems of the earlier systems in that the systems are organized in a modular fashion, with a coherent design philosophy which hides the individual subsystem data-structures within objects.

Feri et al [30] describe a blackboard system, based on a single-threaded geometric reasoner controlling a hierarchical system of knowledge sources (objects) in bottom-up fashion. It identifies 3D objects from data fusion of monochrome and infra-red images, based on prior knowledge of geometric constraints. Their system is able to generate low level object descriptions such as CAD models, but makes no higher level interpretations.

SIGMA developed by Matsuyama and Hwang [45] is a system that finds objects in aerial monochrome views of housing estates. It uses an object-oriented approach, where each object instance is used to establish a concept, and the control is by a sequential failure-driven reasoner working on the equivalents of Horn clauses (in fact the same control mechanism as that used in Prolog). The low-level processing is under the control of the higher-level reasoning. However, the system does not go beyond identifying objects in the scene to include, for instance, high-level concepts like "what kind of housing estate?", or "what are the best utility routings?".

Bell and Pau [7] developed an object-oriented logic programming system for picture interpretation. Their system is based on the Prolog failure-driven/backtracking hypothesize-and-test control mechanism. The object oriented component of the system is implemented with a preprocessor that translates the code into standard Prolog. This three level (feature, application dependent, and object identification) system is used to find objects (cars) in natural scenes.

Although object-oriented, these systems are still single-threaded, and, in fact, sequential (as opposed to concurrent). This imposes restrictions on the range of problems they can easily deal with. In Section 3.1 we explore an alternative to this approach.

2.4   Other Recent Systems

Two interesting recent systems that do not fall into the object oriented paradigm are the following:

Bobick and Bolles [9] introduced a bottom-up system based on a representation space which is a lattice of descriptions, from local image regions to more general descriptions. This includes multiple views as well as refinement and augmentation of the description. This representation scheme is used in a real time system for robot vehicle vision. Sequential images allow improving resolution as the robot moves through the landscape. They eliminate processing artifacts through detecting temporal stability of objects through the image sequence.

Huang et al[41] use a Bayesian belief network and inference engine (HUGIN [3]) in sequences of highway traffic scenes to produce high-level concepts like "car changing lane" and "car stalled". In general, belief networks propagate values around the network as vectors, with each link having associated matrices reflecting the conditional probabilities [52]. For this reason, Huang et al's system is regarded as bottom-up, having a lot in common with the connectionist paradigm. One problem with Bayesian inference is that each node must have a set of exhaustive and mutually exclusive states, which is often difficult to obtain in vision. This problem is further explored in Section 7.4.1.

Tables 2.1 and 2.2 summarize the survey of vision systems given above. The first table is divided into sections (as in the text) corresponding to systems based on ATN, semantic net, constraint propagation and the German group of systems. The second table, corresponding to systems published in the 1990s, is divided into object-oriented systems and the rest.

Paper	Input	Method	Output	Remark
Tropf 1983 [64]	images of objects	ATN	objects	single thr. LL
Bajcsy 1985 [5]	aerial stereo pairs	ATN	scene model via user queries	single thr. LL
Nieman 1985 [51]	images of human heart sequence	production rules over semantic net controlled by A* search	diagnostic descriptions	multi thr. HL
Dellepiane 1987 [23]	sets of slices from brain tomography	prod. rules over semantic net with hyp. & test	brain components	single thr. LL
Govindaraju 1990 [33]	newspaper images	semantic net relates language to vision using hyp. & test.	relates faces to caption text	multi thr. HL
Mulder 1987 [49]	sketch map encoded by plotter cmds	local constraint propagation over graphs	labeled map	multi thr. LL
Boddington 1990 [10]	images of cars	constraint propagation and ATMS	found cars	multi thr. LL
Schirra 1987 [61]	video traffic sequences	user query driven geometric reasoner	traffic descriptions	single thr. HL
Andre 1988 [4]	video sequences of soccer	bottom-up event recognition with transition net	description of soccer match	multi thr. HL BU
Neumann 1989 [50]	video traffic sequences	hyp. & test over semantic net	descriptions of traffic events	single thr. HL.

Table 2.1: *Survey of status of vision systems, pre 1990. Single thr. means single-threaded, referring to systems that process no more than one hypothesis at any instant, as opposed to multi thr.. LL means low level output, HL means high level output. All systems are top-down except those labeled BU (bottom up). Note: for brevity, only the first author on each paper is named.*

Paper	Input	Method	Output	Remark
Draper 1989 [27]	natural images	concurrent processes and blackboard	region labels	multi thr. LL OO (Sec. 3.2)
Feri 1990 [30]	visual & IR images	bottom-up geometric reasoning	region labels	single thr. LL BU OO
Matsuyama 1990 [45]	aerial views of housing estates	hyp. & test over Horn clauses	objects	single thr. LL OO
Bell 1992 [7]	car images	OO Prolog extension	objects	single thr. LL OO
Bobick 1992 [9]	real-time robot image sequence	lattice of desc'n in finite state machine, bottom up processing	objects	single thr. LL BU
Huang 1994 [41]	highway traffic sequences	Bayesian belief inference engine	traffic situation desc'ns	multi thr. HL BU

Table 2.2: *Survey of status of vision systems, post 1990. Single thr. means single-threaded, referring to systems that process no more than one hypothesis at any instant, as opposed to* multi thr.*. LL means low level output, HL means high level output. All systems are top-down except those labeled BU (bottom up). OO means object-oriented architecture. Note: for brevity, only the first author on each paper is named.*

Chapter 3

Foundations

In this chapter, we consider the foundations of some of the ideas used in the design of our system SOO-PIN. We show how the three important strands of: frames, logic programming and object orientation come together in the language Parlog++, and how this is the optimal vehicle for our network-of-frames.

3.1 Human Conceptualization

The systems described in the previous chapter, including the object oriented ones, all have a rather monolithic design based on the sequential processing paradigm. While this is legitimate, it does not capture the ways in which humans describe images, therefore alternative design philosophies are required. From a psycholinguistics viewpoint, Lakoff[43] has pointed out that human categorization cannot be defined in terms of "necessary and sufficient conditions" (which we take to mean a conjunction of propositions), but rather is explained in terms of "idealized cognitive models" (ICM). ICMs are active structures (processes) that define categories in terms of prototype effects, ie, objects are members of categories to the extent that they relate to the prototype via the following:

- "frame-like" structures,

- "image-schematic" structures, which map between spatial concepts having similar schemas, (for example, the container schema refers not only to a cake in an oven and tea in a cup, but also to categories and members),

- metaphor, or structural similarity,

- metonymy, or similarity of salient features or parts.

Lakoff's ideas bear a striking resemblance to those of Herskovits [36], who investigated the spatial prepositions *at, in* and *on* in English, and how they were used. She went on to look at "projective prepositions" which relate one object to another in space. She claims that knowing the objects and their coordinates is not sufficient to select the appropriate preposition. Spatial prepositions have "ideal meanings" which are qualified in the real world by "sense shift" and "tolerance shift". Sense shifts are discontinuous shifts to another, conceptually close relation. For instance, the word *on*, whose ideal meaning is that of *support from below and contiguity*, when used in the expression "the apple on the branch", has been sense-shifted to mean *support from* **above** *and contiguity*. Tolerance shifts involve gradual measurable deviations. For instance in the expression "the book is on the table" when there is a table-cloth between the book and the table [36].

The ways in which objects relate to their ideal meaning, according to Herskovits, is via the following:

- **salience** - referring to the part or aspect of an object that is relevant in the context. For instance, "the queue at the counter" refers to the salient part of the queue, namely the head of the queue, being next to the counter rather than any other part of the queue.

- **relevance** - aspect of relation to be emphasized in the context. For instance, the expression "the light bulb is in the socket" uses the word "in" to emphasize the bulbs function, ie, whether it will work, rather than just its position, where the word "under" would have been just as good.

- **tolerance** - how much deviation from the implied ideal varies with context.

- **typicality** - similar to default reasoning. For instance, the expression "the fountain is behind the town hall" implies that the fountain is near to or next to the town hall, but the ideal meaning of "behind" does not involve proximity, proximity is the default assumption.

Lakoff and Herskovits' ideas suggest the need for mediating active and intelligent agents operating between concepts in the system and the world. Below we explore one scheme, namely frames, mentioned by Lakoff, for implementing such agents.

3.2   Frames and Related Ideas

The notion of frame-like structures emerged in the mid 1970s and was crystallized by Minsky[47] in his 1975 essay "A Framework for Representing Knowledge". In this essay he outlined an approach to the fields of both vision and natural language processing based on the notion of a "frame", which is a data structure for representing a stereotyped situation. Attached to each frame are "terminals" (or slots) in which specific information about the frame is stored. Terminals can contain "demons" for calculating and validating information, or default values, or links to other frames. If sufficient terminals find mismatched data, the frame may invoke another with better matching terminal constraints. Thus, in a vision example, when entering an unfamiliar room, the room frame terminals contain default values like left wall, right wall, ceiling but with no details. The terminal demon for, say, the left wall actively seeks out information to instantiate its values and, if necessary, activates the wall subframe. If the wall demons cannot find good values, then the "room" frame may be replaced by, say, a "backyard" frame. This system has elements of top-down and bottom-up control: bottom-up when a new frame is activated or from the tension of a mismatched frame, top-down when a frame activates its terminal demons.

Similar ideas put forward around the same time are Carl Hewitt's "actors" and Roger Schank's "scripts". An actor[37][38][70] is a potentially active piece of knowledge (procedure) which is activated when it is sent a message. Actors interact by sending messages to other actors, which can take place concurrently. An actor system is defined by the following:

- what actors (objects) exist,

- what messages they receive,

- what they do upon receiving a message,

- what acquaintances (other known actors) each actor has.

Scripts[57][60] are stereotyped sequences of events in a particular context. Scripts consist of a set of slots which have indications of the kind of data within them and possible default values. Scripts differ from frames insofar as they refer to specific events and the slots have specific information in them common to such events, for instance, "entry conditions", "results", "props", "roles" and "scenes". This formalism is based on Schank's earlier work on "conceptual dependency" – the idea that most events portrayed in natural language can be expressed as graphs made up of a small number of primitive acts with "dependencies" between them.

Frames, and the related ideas, were taken up by the natural language processing community, but, until recently, vision applications have been limited. However, a recent example in vision is SCHEMA, developed by Draper et al[27]. In this system there are concurrent processes communicating through a blackboard and the control is frame-based with structure matching and forward chaining. This system is used to segment natural 2D images, but is not concerned with high-level interpretation. It should be pointed out that in such systems one needs to keep the interconnectivity between frames reasonably low, or the time spent handling non-productive messages becomes untenable as the cardinality of the system goes up. Blackboard systems are subject to this, unless one introduces some kind of partitioning of the blackboard. However, in this case, the blackboard system is then logically equivalent to a network-of-frames with appropriate connectivity.

The systems of Bell and Pau[7] and SIGMA[45], do not qualify as networks-of-frames because failure-driven backtracking is incompatible with such an approach. That is, frames, as we define them, are message passing processes that exist in real time and cannot directly retract (ie, backtrack) a message once sent. For instance, if two modules A and B are in communication, and module A experiences a failure and backtracks to some previous decision point, and if, in the meantime, it has sent a message to module B, then in general the system state cannot be reverted to the state which existed at the time of the decision point unless the effects of the message on B were also backtracked. However, if this mechanism existed, then the independence of the modules would be compromised.

In this thesis we develop an alternate implementation of Minsky's frames idea, involving a network of concurrent processes, each dealing with one concept or aspect of the interpretation. One novelty of this system is that the network is extended beyond the object recognition stage to higher-level concepts based on interactions between objects and rules (knowledge, road laws, etc) – for instance "car A gives way to car B due to *road rule X*". This is comparable to the level of output from Neumann's[50] system which generates concepts like "car A is overtaking truck B". However, in Neumann, reported interactions between cars is limited to their spatial relationships and do not refer to road rules or the drivers' intentions.

Another novelty in this system is the use of an object-oriented concurrent logic programming language for its implementation. This environment offers several advantages:

- the high-level declarative nature of logic programming allowed a fast prototyping of the system. It also provides symbolic processing, and pattern matching.

- object-orientation reflects, to some extent, the active and modular nature of human concepts, as Lakoff[43] pointed out, and eases the construction and modification of the system.

- concurrency allows a full and natural implementation of the frames concept. It also positions this system to take advantage of parallel machine architectures.

3.3 Logic Programming

Logic programming (LP) effectively began with Robinson's [59] discovery of the resolution principle. This has since evolved through systems such as Planner[39] to its best known manifestation, Prolog. This is a language based on Horn clause logic (which is a subset of first order predicate calculus) and the resolution principle with a failure-driven backtracking control mechanism. However, Prolog does not exhaust the possible logic programming paradigms. LP languages have different interpretations. For instance, Prolog can be interpreted declaratively, in which the program is regarded as a logical statement, and its meaning is the set of ground formulae that its clauses imply; or procedurally, in which the program is regarded as a precise specification of how the inference will be controlled through time. The procedural interpretation of Prolog is single-threaded.

There is another class of LP, concurrent LP, where the language is a collection of Horn clauses together with the parallel evaluation of a goal with respect to these clauses. This class has a third interpretation based on the idea that goals can be regarded as a system of *concurrent processes*. Concurrent Prolog and Parlog[17] are such languages.

Parlog differs from (sequential) Prolog in how clauses are chosen for execution, Prolog executes each clause in a goal sequentially, backtracking and trying the next clause upon failure. This is the basis of its depth-first search. Parlog has a construct called the "commit" operator which delimits a "guard" condition in a clause – since all the clauses of a goal can be tried concurrently, the first clause for which the guard succeeds is committed to (ie, that clause is selected for evaluation, the rest are ignored), with no backtracking. This is called "don't care" or "committed-choice" non-determinism as opposed to Prolog's "don't know" non-determinism. This mechanism in which alternate clauses are tried concurrently is an example of "or-parallelism". Another concurrency mechanism available in Parlog is "and-parallelism", in which, when a goal is dependent upon the conjunction of several subgoals, the subgoals are evaluated concurrently.

To further elaborate the process interpretation of LP, below we relate process concepts to the corresponding logic programming concepts:

- a long-lived process corresponds to a goal which calls itself recursively.

- A process state corresponds to an unshared variable that a recursive goal passes to itself as an argument upon the recursive call (and thus propagates values from call to call through time).

- Two processes in communication via a channel can be modeled by a conjunction of such recursive goals sharing a variable (which is itself a recursive data structure - ie, a list). For instance, one process, called the producer, instantiates the head of the list to a value, then calls itself recursively with the rest of the list (called the tail) as an argument. The other process, called the consumer, waits for the head of the list to become instantiated. When it is instantiated, the consumer processes the head and then, like the producer, calls itself recursively with the tail as an argument. See Figure 3.1.

- Reply to a query (back channel communication) corresponds to the producer sending an uninstantiated variable to the consumer through the channel. Upon receiving the variable, the consumer instantiates it to a value which is thus available to the producer.

In this way many of the properties of concurrent programming can be mapped into Parlog[58].

3.4 Object Orientation

The object-oriented programming (OOP) paradigm is a model for programming based on the notion of computational objects which communicate via message passing. This paradigm is similar to Minsky's[47] frames and Hewitt's[38] actors described above (in fact Hewitt was influenced by an early version of SMALLTALK). OOP is characterized by the following features:

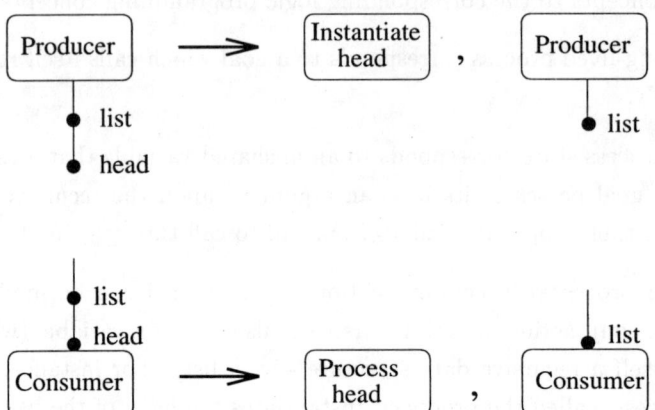

Figure 3.1: The communication channel in Parlog implemented using a shared list. On the left the producer and consumer goals are shown, with their shared variable – the list with its head – as the vertical line with the black disks. On the right is shown the result of the evaluation of the two goals, with the producer instantiating the head to a value to be communicated and then calling itself with the remainder of the list, and the consumer processing the instantiated head and also calling itself with the remainder of the list. This diagram has been adapted from [58].

- creating an object

- sending and receiving messages

- modifying an object's state

- forming class-superclass hierarchies

Shapiro and Takeuchi[62] have shown how these features can naturally be implemented in Concurrent Prolog through the following identifications:

- An object is a process that calls itself recursively and holds its internal state in unshared arguments.

- Objects communicate with each other by instantiating shared variables.

- An object becomes active when it receives a message, otherwise it is suspended.

- An object can respond to a message either by sending the answer back in another message, or instantiating an uninstantiated answer variable in the message.

- Class hierarchies and inheritance can be formed using filters (Shapiro and Takeuchi [62]), which actively pass messages from a class to the superclass when the method to handle a message is not available in the class.

Parlog++[21] is an object-oriented extension to Parlog. The OOP features (except for hierarchies and inheritance) described by Shapiro and Takeuchi are built into the language so that the programmer can write object-oriented code in a natural manner. As Shapiro and Takeuchi point out, their view of objects is based on Hewitt's actor model of computation, which in turn is influenced by and influences Minsky's frames model. Thus Parlog++ is the most suitable language in which to implement our network-of-frames system since it embodies the concurrent frame-based concepts required whilst being a high level logic programming language with the consequent symbolic processing and pattern matching capability.

Chapter 4

Architecture of SOO-PIN

In the last chapter, it was argued that the ideal image interpretation system is a concurrent frame based network. It was also argued that a suitable language in which to implement such a system is the Parlog++ language. What is now required is a system concept and architecture. Hewitt's [38] prescription for defining an actor system (pg. 17), provides a guide for describing SOO-PIN, as explained below. His first point, "what actors exist", is now dealt with.

What is envisaged is a network of independent frames (which are here called "concept-frames"), each dealing with an aspect of the overall interpretation. To this end they communicate with other concept-frames, and also access the (processed) image itself. Concept-frames store instances of the concept it is concerned with (which are called "concept-instances") as data (as opposed to some object-oriented systems where instances are themselves objects or classes). Each concept-frame carries its code (or *methods*, in OO language) which determines its behavior.

The next two decisions in Hewitt's list, "what messages the actors receive" and "what they do upon receiving a message", are now dealt with. The problem of what messages to convey around a network is similar to the situation dealt with by Green [34] in the distributed artificial intelligence (DAI) context. He deals with real-time systems in the area of robotics, and his system, like SOO-PIN, also partly derives from the work of Minsky and Hewitt. In his system, activation framework objects (AFO) form a community of experts

24

which communicate by means of message exchange. These messages come in 3 main types:

- automatic: if the AFO is given or deduces some new piece of information, it will inform those AFOs that it knows and can use this information.

- on-demand: one AFO can ask another about the current value of an hypothesis, or can ask for the AFO to evaluate some data it is sent, or data that it needs to acquire to respond to the request.

- suggestive: in this mode, one AFO sends a message to another that results in a change in the activation level of the evidence for or against some hypothesis that is in the domain of the second AFO.

He points out that automatic messages correspond to forward chaining (or bottom-up), and on-demand to backward chaining (or top-down).

In accordance with Green, SOO-PIN has two classes of message, query and informative, which correspond to Green's on-demand and automatic message types respectively. The query messages contain queries like "list all concept-instances of a certain concept-frame with a certain property" (called in SOO-PIN an `anyInst` message), or "does a certain concept-instance have a certain property" (the `getVal` message). The informative messages, in bottom-up mode, inform concept-frames about the existence of a concept-instance in another concept-frame, and suggests the recipient check specific relations regarding it (the `check` and `create` messages)[1]. The other informative message simply updates specific concept-instances with new properties, and do not initiate any other activity (the `updVal` message). SOO-PIN has no analogy to Green's "suggestive" class of messages, as this relates to the scheduling mechanism used there, although it does bear some resemblance to the `belUpd` message introduced later in Section 7.4.2.3.

---

[1]It is worth noting that the informative messages have the effect of a spreading activation through the network. This is analogous to the mechanism used in the system GRANT (Cohen et al [53]) in which funding agencies are matched with funding applicants.

The final prescription in Hewitt's list – "what acquaintances each actor has", is reflected in SOO-PIN by the list in each concept-frame of other concept-frames that are sent information in automatic (bottom-up) mode. However, SOO-PIN allows concept-frames to react to messages from anywhere.

4.1   Procedural Description of SOO-PIN

In accord with the prescription given above, we now describe how SOO-PIN behaves for a sample case in which the system recognizes a bicycle from low level data.

When a concept-instance of the wheel concept-frame (see Figure 4.1) is created –for whatever reason– it sends messages checkA to all associated concept-frames (according to how the system has been connected to reflect the domain knowledge, as per Hewitt's acquaintance list) informing them of its existence. These messages, in turn, prompt receiving concept-frames (ie, the bicycle concept-frame) to check their "existence criteria", possibly sending inquiry messages inquiryB to other concept-frames to establish the criteria, and, if successful, they create a concept-instance (ie, a bicycle instance). This new creation, again, results in checkB messages being sent, for instance, to the street concept-frame, and thus completing the loop. The existence criteria, and messages sent as a result of existence, for all the interconnected concept-frames in the system, constitute the "World Model" (as Bajcsy et al [5] call it) – meaning the prior knowledge about the situation that gives meaning to the low-level data in a given image. The specific state of the system as a result of the low level data – the "Scene Model" [5] – is embodied in the list of instances for each concept-frame.

As pointed out above, this architecture embodies both data-driven and knowledge-based control, where the automatic (**check** and **create**) messages represent the data-driven mode, and the on-demand (**getVal** and **anyInst**) messages represent the knowledge-based mode, in that these messages are prompted from the exploration of higher-level hypotheses. Using these two modes together avoids the large search space that results from using either alone. For instance, knowledge-based control can entail exhaustive depth-first search as the system begins searching at the top nodes of the search tree, and must check all nodes down to the data level, usually with failure-driven backtracking. On the other hand, data-driven search modes entail combining all data in all possible patterns in order to explore higher nodes in the search tree, and as Tsotsos [65] has pointed out in the area of vision, this is NP-complete (meaning that it has exponential complexity). Our claim that mixing top-down and bottom-up modes is more efficient than either alone, is backed up in the area of linguistic parsing by Allen [2], who gives mixed-mode chart parsing as an example.

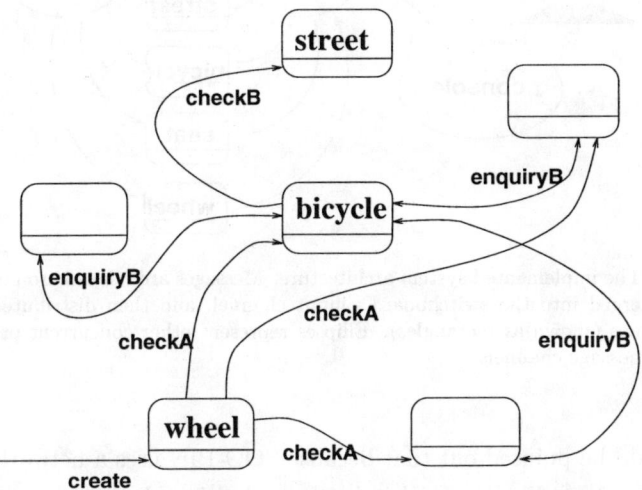

Figure 4.1: The SOO-PIN system concept: object concept-instance creations activate other associated objects through sending "check" messages. These activations in turn prompt further activations.

SOO-PIN requires a mechanism whereby a number of concept-frames can communicate with each other. One means by which this can be arranged, using a mechanism pointed out by Ringwood [58], is to use a central process with which all concept-frames are in a client-server relationship. The output

channels from all these clients (the concept-frames) can then be merged together and input into the server. In SOO-PIN, this server process is called
the "switchboard". Thus the function of the switchboard is to keep track of
the concept-frames' input channels and redirect messages from one concept-
frames to another (see Figure 4.2). It is also the switchboard's job to spawn
off new concept-frames as required. In Parlog++, spawning takes place
through invoking goals using and-parallelism, as mentioned in Section 3.3.
The concept-frames and switchboard also write output messages directly to
windows on the user console.

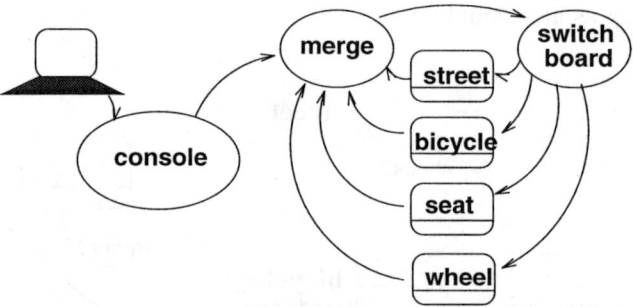

Figure 4.2: The implemented system architecture. Messages are relayed around the system
by being merged into the switchboard's input channel, and then distributed to various
concept-frames (shown as rectangles). Ellipses represent other concurrent processes, and
arrows are message channels.

It should be pointed out that because SOO-PIN uses a network-of-frames
for its control-mechanism, and is implemented in a logic programming language, it falls into Bunke's [12] hybrid category of PC-SN (Predicate Calculus
– Semantic Net) in his overview of pattern recognition architectures.

Having sketched the structure of SOO-PIN, it is now necessary to fill it
out by considering some of the other constraints and requirements of the
system. One of the prime considerations with SOO-PIN is that it is not
simply a logic programming symbolic inference engine, but deals specifically
with **image interpretation**. Hence the following section is concerned with

how spatial data is integrated into SOO-PIN. This theme is taken further in the following section (Section 4.3) when we consider the interplay between the spatial data of SOO-PIN and high-level concept-frames, especially those of compound objects.

4.2   Spatial Data

If a compound object C is composed of objects A and B, then an object D may be near C but not near A or B. That is, "nearness" depends on the *scale* of the objects involved. This is not something that can be handled purely symbolically, but must refer to the spatial data underlying the objects A, B and D. Thus, when the system needs to instantiate the nearness predicate for D and C, it first needs to determine the components of C, and then the components' spatial coordinates. If there is a deeply nested set of components the system would need to search down the component tree for the components with spatial data. This could be a large search and adds to the complexity of the system. For this reason, in SOO-PIN, when a compound object is created, an entry is inserted in the spatial database with the spatial coordinates of the new object. These entries are calculated from the coordinates of its components, and are used to calculate spatial predicates for compound objects in the same way as for primitive objects.

Of course, the nearness predicate is but one of many potential spatial predicates and, in general, the system could not store a large range of spatial relationships as symbolic predicates. For instance, the system may store the fact that objects A and B are near, but also requires to compute their collinearity. In fact, as Herskovits [36] has pointed out, even the same predicate can vary in meaning depending on the context – illustrating the need to refer back to the spatial data. This is an important principle in the area of computer vision (and other areas of AI as well), and relates to the necessity of embedding cognitive systems in their environment in order to give meaning to their symbols – situatedness. Situationists have made radical claims that cognitive systems *only* use context for their representations, and that there

are no internal representations [16]. However, Slezak [63] has pointed out the
need to distinguish the representation used internally to implement cognitive
systems from that used for external communication, and how this clarifies
some of the situationists' claims. However, in the context of SOO-PIN, this
reduces to noting the distinction between the representation of the spatial
data and the symbolic language of Parlog++.

4.3   Compound Objects

High level concept-frames (for instance, compound objects) introduce an
"identity" problem. When a new concept-instance is to be created, the system
must check through existing concept-instances to see if the instance already
exists. The question of what constitutes a "unique key" to the concept-
instance can be subtle, and depends upon the domain (prior) knowledge of
the world. To further elaborate, if a system determines that a car exists at
some point in space, but there is already another car there, the knowledge
that only one car can exist at any given point would lead the system to identify the two. However, other objects can coexist in space, for instance, cakes
and ovens. This question of how ordinary objects work and interact has been
discussed by Hayes [35]. In SOO-PIN, each concept-frame has its domain
knowledge built in. Thus the "existence criteria" of concept-frames embody
the question of the identity of concept-instances.

Domain knowledge also comes into play in the case where "negative information" is generated, for instance, from the deletion of a concept-instance.
Here the receiving concept-frame applies its domain knowledge to decide
whether any concept-instance critically depends upon the sender. If so, it
too is deleted. However, the situation becomes complex when evidence from
several sources justify the existence of the concept-instance. Deletion hap-
pens when, for instance, a bicycle is found to be joined to a third wheel (in
the case of a tricycle), or an existing component wheel is deleted (making
it a unicycle). This is an example of "non-monotonic reasoning" [46]. It
is not compatible with traditional predicate calculus, but in logic program-

ming systems can be implemented procedurally. Humans use non-monotonic reasoning often, for instance, when a default assumption is over-ridden by subsequent information. Thus, in general, frame-based systems also deal in non-monotonic reasoning because they too use slots with default values.

4.4   Concept-Frame Structure

This section defines the syntax of the data structures used in SOO-PIN concept-frames and instances, and in the messages.

The concept-frame needs to store the concept-instances together with their properties. The mechanism chosen to do this is the list, a data structure used extensively in Parlog++, for instance, in implementing the data channel (see Figure 3.1). Thus the concept-frame maintains a list of concept-instances, which have the following structure:

```
inst(Id,PropList)
```

Here, each concept-instance has its identity string, Id above, and maintains a list of its properties, be they unary or binary, as the PropList list. These structures are further broken down in the following table:

Id	= id(ObjectType,InstNumber)
ObjectType	is the same as the concept-frame name
InstNumber	is a unique identifier for the concept-instance
PropList	is a list of Prop
Prop	= reln(RelnType,OtherId) or desc(Desc)
RelnType	is any relationship, ie ''near'', ''joined'', ''above''
OtherId	is the Id of the other party to the relationship
Desc	is a structure giving a description, ie ''colour(green)''

Below is a schematic example of a typical concept-frame, showing its reaction to received messages. The code can be read, between the "clauses" and "end" statements, as a simple case statement (as in the language C), where the system, upon receiving a message, searches down the list of clause heads for one that matches the message. The clause body is then evaluated (not shown here).

bicycle.
```
            Out o-channel        %output channel name
invisible InstList state, OutFile state
                        %state variables

initial open(bicycle.log,write,OutFile)
                        %activities when starting up concept-frame
clauses
            create(Id,PropList)
                        %Concept-instance Id is added to InstList
                        %with its Props

            check(composedOf)
                        %checks if a concept-instance should be created

            check(Prop)
                        %checks if a concept-instance has relationship Prop
                        %with sender.

            negCheck
                        %sender has been deleted,
                        %checks if a concept-instance should be deleted
                        %in turn.

            inquiry(Id,Prop,FoundProps)
                        %Prop is searched for in concept-instance
                        %Id, and the answer bound to FoundProps

            update(Id,Props)
                        %Props are added to property list of
                        %concept-instance Id

            WrongMsg writeMy(OutFile,[WrongMsg]).
                        %default clause for erroneous messages

            last
                        %clause to execute upon input channel close

    end.     %end of concept-frame
```

The first message received by the concept-frame activates the "initial" clause, the persistent data of the concept-frame is stored in the "state" variables defined in the "invisible" section, output messages to other concept-frames are placed in the "o-channel" declared after the concept-frame name, user messages are written to the text file defined in the initial section. Erroneous messages fall through to "WrongMsg" where they generate error text messages, and finally, when the input channel closes, the "last" clause is executed.

In accordance with the parallelism of Parlog++, all the clauses in the above code can execute in parallel, spawning a process for each message on the input channel. They can also run sequentially, finishing one message before starting the next. In SOO-PIN, the concept-frames run in parallel with each other, while within a concept-frame, the code is implemented in parallel for simple situations like sorting a list. Everything else is executed sequentially.

Table 4.1 shows the action initiated by each message. This varies with the logic built into each concept-frame, but these are the overall consistencies.

```
MESSAGE
        ACTION
              RESULTING MESSAGES
create(Id)
        creates a concept-frame if it does not already exist
                check message sent to associated concept-frames


check(composedOf)
        checks if concept-instance already exists in relation
        composedOf with sender.
                If so, update sent to sender with this fact.
                If not, then sends inquiry to associated
                concept-frames to satisfy existence criteria.
        checks if sender is a new component of existing instance
                update sent to sender with this fact.
        checks if this is a new concept-instance,
                check sent to associates of new instance,
                update sent to sender with "composedOf" Prop.
        checks if any concept-instance should be deleted as a
        result of this information,
                negCheck sent to associates of deleted instance.


check(Prop)
        checks if any concept-instance exists in rel'n Prop with sender
                sends inquiry messages to determine this.


negCheck(Id)
        informs receiver that Id has been deleted, may result
        in deletion of this concept-frame.
                If so, negCheck sent to associates.


inquiry(FoundList,Prop)
        returns in FoundList a list of instances of this
        concept-frame satisfying Prop, or list of
        properties of specific instance satisfying Prop.


updval(Id,Prop)
        updates Id with Prop properties.
```

Table 4.1: *Table showing the actions performed, and the messages sent, by a concept-frame as a result of receiving various messages. The received messages are shown in the left column, the actions are given in the middle column, and the resulting messages in the right column.*

Chapter 5

Simple Scenarios

Having described the motivation and structure for SOO-PIN, we now consider two simple implementations which demonstrate its power. First we deal with the "cutlery" scenario, which uses no spatial database but calls upon pre-defined relational predicates to drive the network. Second, we describe the "wheels" scenario, which extends the implementation to include "negative" information (see Section 4.3) in a more complex network.

5.1 Cutlery Scenario

Consider a camera observing a tabletop with various objects on it. We also assume that the low-level vision problem of identifying and locating the various objects has been solved, and the job of SOO-PIN is to interpret the image. The scenario (see Figure 5.1) consists of the primitive concept-frames `knife`, `fork`, `spoon` and `chopstick` which are created initially from the low level vision system. The higher-level `stick_set` embodies the concept of a pair of near and parallel chopsticks, `setting` which deals with various combinations of knife, fork and spoon, `chinese_setting` which similarly deals with stick_sets and spoons, and at the top level is `dinner` and `yumcha`, which are interpretations of the scene as a whole. Note that the primitive concept-frames, ie, knife and fork, have no existence criteria as they are created directly from the low-level data.

The spatial database in this scenario consists of various assertions about which objects are near and parallel to each other, which is sufficient to group the various cutlery pieces into "table settings".

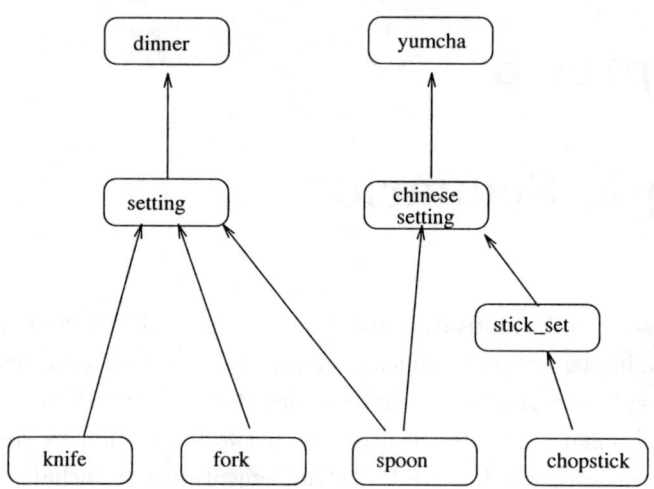

Figure 5.1: *The cutlery scenario network-of-frames. The arrows refer to* check(composed-Of) *and* create *messages, inquiry and update messages are not shown.* stick_set *refers to a pair of chopsticks.*

Creation of the primitive concept-frames (knife, fork, spoon and chopstick) results in the following messages:

```
knife:
        check near to fork
        check near to spoon
        check composedOf to setting

fork:
        check near to knife
        check near to spoon
        check composedOf to setting

spoon:
        check near to knife
        check near to fork
        check near to stick-set
        check composedOf to setting
        check composedOf to chinese-setting

chopstick:
        check near to spoon
        check composedOf to stick-set
```

The next level in complexity is that of compound objects. Defining these concept-frames requires more information, namely, the "existence criteria", the "uniqueness criteria" (ie, does the system add the sender to an existing concept-instance, or create a new one), and the messages sent upon new concept-instance creation.

Note that chopstick checks component-hood only with stick-set. As a single chopstick is not an acceptable implement in the chinese-setting, it is the stick_set that behaves like a cutlery implement.

```
Existence criteria:
        chopstick sending message is joined and
        parallel to another chopstick
Messages sent upon creation:
        check near to spoon
        check composedOf to chinese-setting
```

Setting results from a knife, fork or spoon in certain spatial relationships:

```
Existence criteria:
        sender is near and parallel to other knife, fork
        or spoon.
Uniqueness criteria:
        sender is not near and parallel to any cutlery
        already in a setting.
Messages sent upon creation:
        check composedOf to dinner.
```

Chinese-setting results from a stick-set and a spoon being in a certain spatial relationship:

```
Existence criteria:
        stick-set sending check message is near and
        parallel to spoon.
        spoon sending check message is near and
        parallel to stick_set.
Uniqueness criteria:
        sender is not near and parallel to any cutlery
        already in a setting.
Messages sent upon creation:
        check composedOf to yumcha.
```

Dinner is deduced from the activation of one or more settings:

```
Existence criteria:
        Any setting passed.
Uniqueness criteria:
        Any subsequent settings are added to the first dinner created.
No messages are sent upon creation as dinner is a top level production.
```

Yumcha is deduced from the activation of one or more chinese-settings:

```
Existence criteria:
        Any chinese-setting passed.
Uniqueness criteria:
        Any subsequent chinese-settings are added to the first yumcha
        created.
No messages are sent upon creation as yumcha is a top level production.
```

5.2   Trial Runs of Cutlery Scenario

This first simple system was initialized with:

```
create(id(knife,1))
create(id(fork,3))
```

and the "spatial database" was loaded with the following predicates:

```
near knife 1 fork 3
near fork 3 knife 1
parallel knife 1 fork 3
parallel fork 3 knife 1
```

When the system was run with this input, the following messages were passed
through the switch concept-frame. Note that messages are structured
msg(target,message_body):

```
 1    msg(knife,create(id(knife,1),[]))
 2    msg(fork,create(id(fork,3),[]))
 3    msg(knife,check(reln(near,id(fork,3)),_))
 4    msg(spoon,check(reln(near,id(fork,3)),_))
 5    msg(setting,check(reln(composedOf,id(fork,3)),_))
 6    msg(knife,anyInst(_2362,reln(near,id(fork,3))))
 7    msg(fork,check(reln(near,id(knife,1)),yes))
 8    msg(spoon,check(reln(near,id(knife,1)),_))
 9    msg(setting,check(reln(composedOf,id(knife,1)),_))
10    msg(fork,updVal(id(fork,3),[reln(near,id(knife,1))]))
11    msg(knife,updVal(id(knife,1),[reln(near,id(fork,3))]))
12    msg(knife,getVal(id(knife,1),reln(parallel,id(fork,3)),_))
13    msg(spoon,anyInst(_5847,reln(near,id(fork,3))))
14    msg(fork,updVal(id(fork,3),[reln(partOf,id(setting,4439))]))
15    msg(knife,updVal(id(knife,1),[reln(partOf,id(setting,4439))]))
16    msg(dinner,check(reln(composedOf,id(setting,4439)),_))
17    msg(fork,updVal(id(fork,3),[reln(parallel,id(knife,1))]))
```

As can be seen, the knife and fork sent check messages to spoon and
setting (messages 4,5), spoon had no reaction as there were none, but
setting reacted by sending inquiry messages (anyInst and getVal) to knife
to determine if any concept-instances were near and parallel to each other
(6,12). When setting was successful at creating an instance, it then sent
update (updVal) messages informing its components of their new status
(14,15,17). It then sent a check message on to dinner (16). Since dinner is
created from any setting, it, in turn, generated the following output message:

```
*dinner 1 consists of the following 1 settings
    setting 4439 consists of the following pieces
        fork 3
        knife 1

no spoons, perhaps there is no desert!
```

It can be seen that the system generated a high-level interpretation on the minimal input data by a process of each concept-frame dialoguing with its associates. With this architecture it is quite easy to vary the response of the system, and the high-level concept-frames are particularly relevant for this purpose.

The next trial of the cutlery scenario involved an oriental meal. Here the stick_set was introduced to demonstrate that the system coped with a compound object that required updating the "spatial database", allowing other concept-frames to interact with it in the same way as for a simple object.

The system was initialized with:

```
create(id(spoon,1)
create(id(chopstick,1)
create(id(chopstick,2)
```

and the "spatial database" was loaded with the following predicates:

```
joined chopstick 1 chopstick 2
joined chopstick 2 chopstick 1
parallel chopstick 1 chopstick 2
parallel chopstick 2 chopstick 1
near spoon 1 chopstick 1
near chopstick 1 spoon 1
parallel spoon 1 chopstick 1
parallel chopstick 1 spoon 1
```

When the system was run with this input, the following messages passed through the switch:

```
 1    msg(chopstick,create(id(chopstick,1),[]))
 2    msg(chopstick,create(id(chopstick,2),[]))
 3    msg(spoon,create(id(spoon,1),[]))
 4    msg(stick_set,check(reln(near,id(spoon,1)),yes))
 5    msg(knife,check(reln(near,id(spoon,1)),yes))
 6    msg(fork,check(reln(near,id(spoon,1)),yes))
 7    msg(chinese_setting,check(reln(composedOf,id(spoon,1)),_))
 8    msg(setting,check(reln(composedOf,id(spoon,1)),yes))
 9    msg(fork,anyInst([],reln(near,id(spoon,1))))
10    msg(knife,anyInst([],reln(near,id(spoon,1))))
11    msg(stick_set,anyInst(_10462,reln(near,id(spoon,1))))
12    msg(spoon,check(reln(near,id(chopstick,1)),yes))
13    msg(chopstick,updVal(id(chopstick,1),[reln(near,id(spoon,1))]))
14    msg(stick_set,check(reln(composedOf,id(chopstick,1)),_))
15    msg(spoon,check(reln(near,id(chopstick,2)),_))
16    msg(stick_set,check(reln(composedOf,id(chopstick,2)),_))
17    msg(chopstick,anyInst(_598,reln(joined,id(chopstick,1))))
18    msg(chopstick,getVal(id(chopstick,2),reln(parallel,
      id(chopstick,1)),_))
19    msg(chopstick,updVal(id(chopstick,1),[reln(partOf,
      id(stick_set,4440))]))
20    msg(chopstick,updVal(id(chopstick,2),[reln(partOf,
      id(stick_set,4440))]))
21    msg(spoon,check(reln(near,id(stick_set,4440)),_)),
      id(stick_set,4440)),_))
22    msg(chopstick,updVal(id(chopstick,2),[reln(partOf,
      id(stick_set,4440))]))
23    msg(chopstick,updVal(id(chopstick,1),[reln(parallel,
      id(chopstick,2))]))
24    msg(spoon,anyInst(_3803,reln(near,id(stick_set,4440))))
25    msg(stick_set,updVal(id(stick_set,4440),[reln(near,id(spoon,1))]))
26    msg(spoon,getVal(id(spoon,1),reln(parallel,id(stick_set,4440)),_))
27    msg(stick_set,updVal(id(stick_set,4440),[reln(partOf,
        id(chinese_setting,4441))]))
28    msg(spoon,updVal(id(spoon,1),reln(partOf,id(chinese_setting,4441))))
29    msg(yumcha,check(reln(composedOf,id(chinese_setting,4441)),_))
30    msg(stick_set,updVal(id(stick_set,4440),reln(parallel,id(spoon,1))))
```

It can be seen that amongst a number of messages from the spoon exploring
the dinner option (messages 8,9,10), the stick_set instantiated itself from the
chopsticks (14,16,17,18,19,20,22,23) and determined its proximity and paral-
lelness to the spoon (21,24,25,26). This involved updating the stick_set
entry in the spatial database with the relationships of its components, the
chopsticks. The creation of the stick_set initiated the chinese_setting (27,28)

and thus the yumcha (29), which, at the end of processing, generated the
following story:

```
*yumcha 1 consists of the following 1 settings
    chinese_setting 4441 consists of the following pieces
        stick_set 4440
        spoon 1

a lonely yumcha
```

Of course, with a real spatial database, the entry for stick_set need not simply
have inherited the relationships of the chopsticks, but could have generated
any appropriate entry. This initial prototype uses only symbolic predicates.

In summary, the cutlery scenario demonstrates how the network-of-frames
functions with a simple domain, for instance, the concept of mixed top-down
and bottom-up control, and concurrent execution of concept-frames, as is
evident from the messages passing through the switchboard. Although the
spatial data is here emulated with a symbolic list, this implementation also
demonstrates a compound object (the stick-set) deriving its spatial properties
from its components. In the next section the use of negative information (see
Section 4.3) is explored.

5.3   Wheels Scenario

In this scenario, it is assumed that the scene contains wheeled vehicles, the
job of the interpretation system is to deduce what the image is about on
the basis of what vehicles it finds. Once again, this is a symbolic system
only, in that the input consists of declarations of the primitive objects, and,
again, the spatial data consists of symbolic forms of their spatial relationships.
The vehicles under consideration are the unicycle, bicycle and tricycle. The
primitive concept-frames are the wheel and the seat. The interconnections
corresponding to the world model are shown in Figure 5.2. One reason for
introducing this scenario is to explore the operation of the negCheck message.
This arises from the unicycle concept-frame when it finds a second wheel

attached to a unicycle instance. Another reason is to explore the scaling of the system.

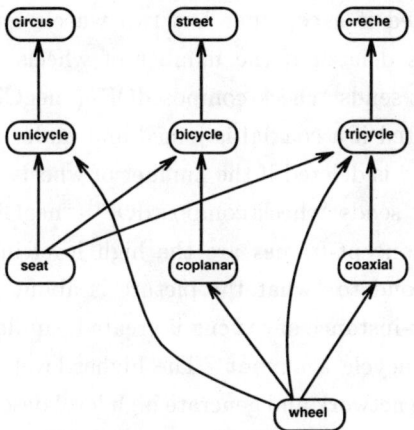

Figure 5.2: *The Wheels Scenario network. Boxes represent concept-frames and arrows show where CHECK(composedOf) messages are sent.*

The starting point is the wheel concept-frame. This is a primitive concept-frame and is created directly. Upon creation, "check joined" messages are sent to seat and wheel, and "check composedOf" messages are sent to unicycle, coplanar, coaxial and tricycle.

The seat concept-frame is similar, it too is a primitive concept-frame, and when created sends "check joined" to wheel, and "check composedOf" to unicycle, bicycle and tricycle.

Coplanar is a compound concept-frame and represents two or more wheels that are joined and coplanar. Its existence criteria are that the wheel passed to it is joined and coplanar with any other wheel. Upon creation it sends "check joined" to seat and "check composedOf" to bicycle, provided there are precisely two wheels.

Coaxial is similar to coplanar, as it represents two wheels that are joined and whose axes are collinear. Upon creation, however, this concept-frame sends "check composedOf" to tricycle.

Unicycle is created if a wheel is joined to, and under, a seat. It is deleted if another wheel is added to the system. Upon creation (deletion) this sends "check composedOf" ("negCheck") to circus.

Bicycle is created if a coplanar has two wheels and is joined to, and under, a seat. It is deleted if the number of wheels exceeds two. Upon creation (deletion) it sends "check composedOf" ("negCheck") to street.

Tricycle is created if a coaxial is joined and under a seat, and is joined to another wheel, and is deleted if the number of wheels exceeds three. Upon creation (deletion) it sends "check composedOf" ("negCheck") to creche.

The remaining concept-frames are the high level interpretations of the system, and correspond to "what the picture is about", ie, if a unicycle is found then a concept-instance of circus is created, similarly a tricycle results in a creche and a bicycle a street. The higher-level concept-frames send inquiries through the network and generate high level descriptions of the scene.

5.4   Trial Runs of Wheels Scenario

In the trial run described below, a unicycle and circus was generated initially. Then a second wheel was added. It can be seen how the system dismantled the circus hypothesis and went on to generate the bicycle/street-scene hypothesis.

Firstly, the system was initialized with:

```
msg(seat,create(id(seat,1),[])),
msg(wheel,create(id(wheel,1),[]))
```

and the database with:

```
below wheel 1 seat 1
above seat 1 wheel 1
joined wheel 1 seat 1
joined seat 1 wheel 1
```

Since many messages passed through the switchboard only a selected output of the concept-frames are shown below:

```
wheel:
        *created id(wheel,1)
        *check relation joined from seat 1
        *update id(wheel,1) with [reln(joined,id(seat,1))]
        *enquiry from id(wheel,1) re joined
        *enquiry from id(seat,1) re joined
        *update id(wheel,1) with [reln(partOf,id(unicycle,4448))]
        *update id(wheel,1) with [reln(below,id(seat,1))]

seat:
        *created id(seat,1)
        *check relation joined from wheel 1
        *update id(seat,1) with [reln(joined,id(wheel,1))]
        *enquiry from id(wheel,1) re joined
        *enquiry to id(seat,1) re reln(above,id(wheel,1))
        *update id(seat,1) with [reln(partOf,id(unicycle,4448))]

unicycle:
        *id(unicycle,4448) created from wheel 1
        *check existing [id(unicycle,4448)] from seat 1

circus:
        *circus 1 created from unicycle 4448
```

At this stage the system had deduced a unicycle from the single seat and wheel, and from the unicycle deduced a circus scene. However, when another wheel joined to the first was introduced, things became more complicated, as shown below:

```
message to wheel:
           msg(wheel,create(id(wheel,2),[]))
```

together with the spatial relations:

```
joined wheel 1 wheel 2
coplanar wheel 1 wheel 2
joined wheel 2 wheel 1
coplanar wheel 2 wheel 1
below wheel 2 seat 1
above seat 1 wheel 2
joined wheel 2 seat 1
joined seat 1 wheel 2
```

the system went on to produce changes as reflected by the output of the following concept-frames:

```
wheel:
        *created id(wheel,2)
        *update id(wheel,2) with [reln(joined,id(seat,1))]
        *enquiry from id(wheel,2) re joined
        *enquiry to id(wheel,1) re reln(coaxial,id(wheel,2))
        *enquiry to id(wheel,1) re reln(coplanar,id(wheel,2))
        *enquiry from id(wheel,2) re joined
        *enquiry from id(seat,1) re joined
        *update id(wheel,2) with [reln(below,id(seat,1))]
        *update id(wheel,2) with [reln(partOf,id(coplanar,4449))]
        *update id(wheel,1) with [reln(partOf,id(coplanar,4449))]
        *update id(wheel,2) with [reln(coplanar,id(wheel,1))]
        *id(unicycle,4448) removed from all relations

coplanar:
        *check relation joined from wheel 1 not found
        *check relation joined from seat 1 not found
        *enquiry from id(seat,1) re joined
        *check relation joined from wheel 2 not found
        *id(coplanar,4449) created from wheel 2
        *update id(coplanar,4449) with [reln(joined,id(seat,1))]
        *update id(coplanar,4449) with [reln(partOf,id(bicycle,4450))]
        *update id(coplanar,4449) with [reln(below,id(seat,1)),
                reln(below,id(seat,1))]

unicycle:
        *deleting instances id(unicycle,4448) composed of id(wheel,1)

circus:
        *deleting instances composed of id(unicycle,4448)

bicycle:
        *id(bicycle,4450) created from coplanar 4449

street:
        *street 1 created from bicycle 4450
```

It can be seen how the unicycle found that the new wheel was joined to its
component wheel, which satisfied its delete criterion, thus it sent off negCheck
messages to its associates. Meanwhile, the new wheel formed the coplanar
compound with the original wheel, consequently the bicycle concept-frame
created an instance based on the coplanar being joined, and under, the seat.
Note how, like the stick_set in the previous scenario, this is an instance of
a compound inheriting spatial relations from its components, in this case the

coplanar inherited the wheels' relations with the seat.

5.5   Summary

This chapter described two experimental implementations of the SOO-PIN concept, the cutlery scenario and the wheels scenario. The first experiment demonstrated, as expected, that the network-of-frames was able to come to the correct conclusions for a simple example. The concept of mixed top-down and bottom-up control, and concurrent execution of concept-frames was shown, and it also demonstrated the "spatial database" [1] being updated with new relationships derived for deduced compound objects.

The second experiment, the wheels scenario, showed a more complex network-of-frames which embodied an extra message, negCheck, to deal with deleted objects.

Based on the same architecture for high-level interpretation, in the following chapter we will discuss a more complex network for real scene interpretation.

---

[1]in this early implementation the spatial database was actually a list of spatial predicates.

Chapter 6

Interpretation of Traffic Scenes

In order to test the capabilities of the SOO-PIN concept, the system is used to interpret real images of outdoor scenes, in a context which is rich enough to require an interesting network-of-frames, and yet constrained enough to be tractable. The context chosen is that of traffic scenes. Here the images are aerial views of city intersections (see Figure 6.1), taken by video. The low-level processing is carried out to extract relevant objects (ie, cars) (see Section 6.5). These objects are made available to the network in the form of a spatial database and initiating messages passed to the low-level (primitive) concept-frames. The traffic scenario was implemented in stages. Initially it was implemented to run on single images, the low level data consisting of labeled cars together with their centroid and major axis coordinates. Later, the traffic scenario was extended to find and utilize the velocities of cars, and the SOO-PIN system was extended to handle uncertainty. This chapter deals with the initial traffic interpretation system.

The task of the network is to interpret vehicle activities (ie, vehicle A is turning right from the west) and to produce analyses of the scene of interest to, for instance, highway engineers and traffic light controllers. For example, the ability to analyze image data for the following queries:

- whether the car is on the wrong side of road or intersection,

- when a car should give way to another. For instance:

48

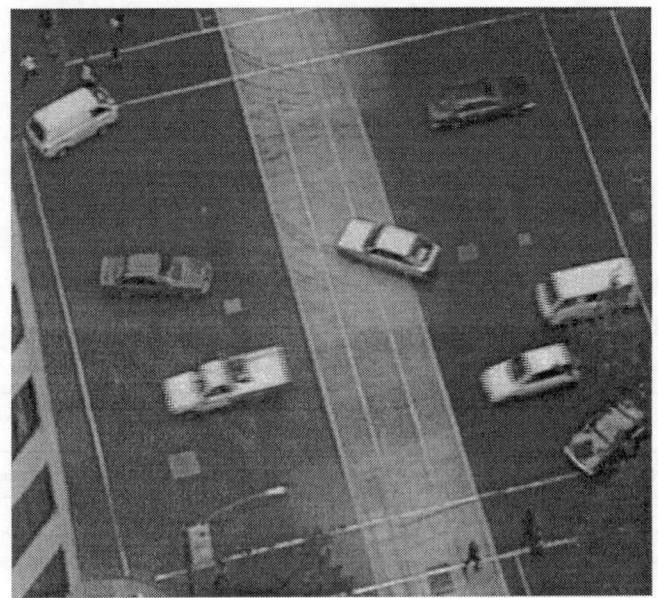

Figure 6.1: *Typical traffic scene processed by the SOO-PIN system*

⬦ give way to right at intersection,

⬦ give way to oncoming when turning right,

⬦ at T-intersections cars in ending road give way to those
on through road,

⬦ at traffic lights or give way signs,

• traffic jams (ie, give-way deadlocks).

These concepts are incorporated in the network-of-frames shown in Figure 6.2,
which is discussed in more detail below.

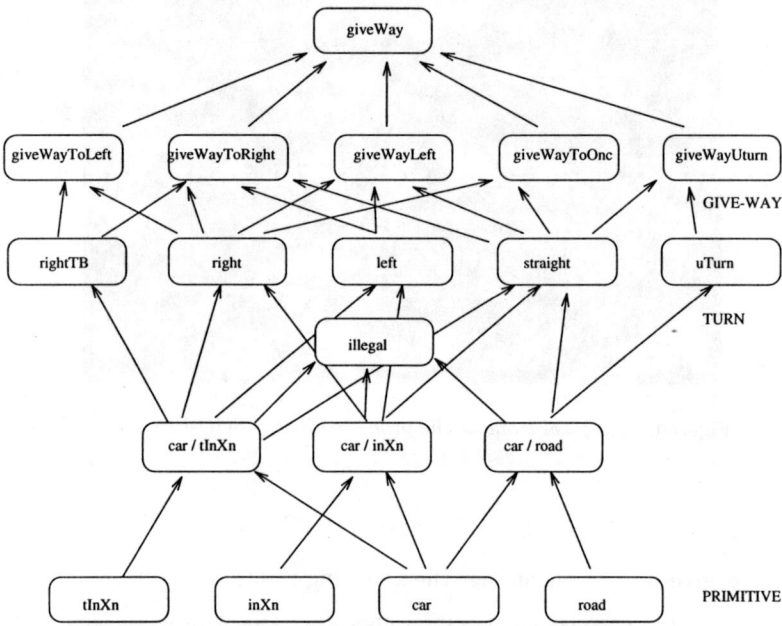

Figure 6.2: *The traffic scenario network-of-frames. The arrows refer to check or create messages, inquiry and update messages are not shown. inXn refers to "intersection", tInXn to "T-intersection", rightTB to a right turn into the through-road of a T-intersection, carInXn to the concept of a car in an intersection, carTInXn refers to a car in a T-intersection, and carRoad to a car in a road.*

6.1 Primitive Concept-Frames

The concept-frames `car`, `inXn`, `tInXn` and `road` correspond to the objects either found by low level processing (see Section 6.5), or given (ie, intersection, T-intersection and road are constant and input manually). These concept-frames simply send `check` messages higher in the network, store information about the concept-instances, and respond to queries about them (in much the same way as the low-level concept-frames described earlier).

6.2 Turn Concept-Frames

On this level, `carInXn`, `carTInXn` and `carRoad` determine the containment of cars in road structures (roads and intersections), ie, "car A is in intersection B". Upon receiving a message from a car or road structure concept-frame, these concept-frames activate a procedural routine written in the C language that read the spatial database and determine whether any of their instances contained the car. This is done because the high-level interactions of a car are dependent on the road structure.

When concept-instances of these concept-frames are created, a call is made to a C routine, `turn.c`, to determine what the car is doing in the road structure. Because images of intersections are, in general, not taken from directly above, this routine first uses an affine transformation that maps the intersection coordinates onto a square to normalize the car positions and heading angles (using transformations derived from [69]). This transformation is also useful because not all intersections are square, and so require to be normalized. The various car activities, ie, right turn, left turn and straight from the north, south, east and west, are unambiguously defined by regions within a 3D product space of the normalized car positions and heading angles (see Figure 6.3). This is analogous to the "spatio-temporal" buffer of Mohnhaupt and Neumann [48] in which traffic events such as turning or overtaking are represented as subsets of the "4D phase-space" of position, velocity direction and speed. The "phase-space" used in SOO-PIN is of lower dimensionality, being simply position and heading. The turn activity found by `turn.c` is sent in a `create` message to the appropriate concept-frame, ie, `right`, `rightTB` (turning right into the through road of a T-intersection), `left`, `straight`, `uTurn` and `illegal`.

Note that heading angle, determined from the major axis of the car, contains no information about which end of the car is the front. This is deduced from within `turn.c` by assuming that cars are not heading in an illegal direction on the road. This problem of determining heading angle is dealt with directly by finding car velocities (see Section 8.1).

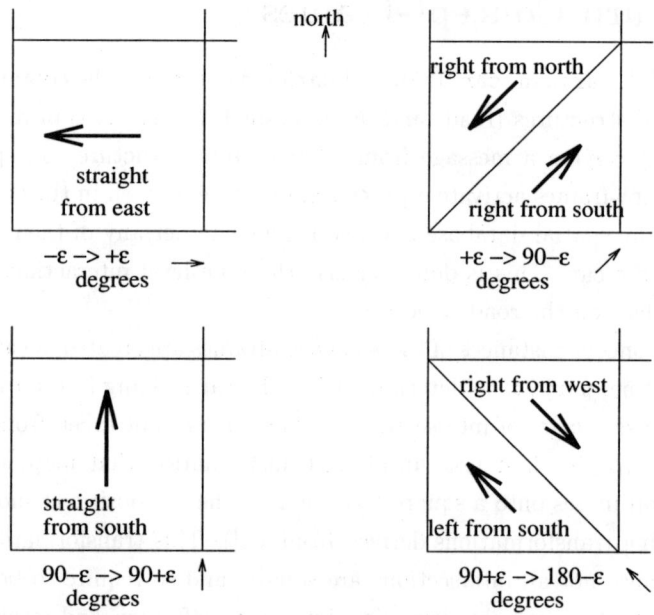

Figure 6.3: *Diagram showing how* `turn.c` *determines car activity from car position and angle. Given that the car is in the south-west corner of the intersection, for the angle range below each square, the system returns the activity as shown within the regions. The other 3 corners of the intersection work similarly.* ε *is the maximum deviation from straight-ahead that is accepted as straight, 6 degrees is used in the system.*

Note that `turn` concept-frames have no existence criteria and so instances are created by fiat. However, these are still useful concept-frames as they are a repository of information about what the various cars are doing, and can be accessed by the usual inquiry messages from other concept-frames.

6.3 Give-Way Concept-Frames

When turn concept-instances are created, they send `check` messages to the appropriate give-way concept-frames. These concept-frames check the context of the car given in the message to see if any other vehicle is in a give-way relationship with it. For instance, if the `right` concept-frame sends a check message to the `giveWayToOnc` (give way to oncoming concept-frame, this first determines the car's heading, then checks the `straight` concept-frame for any cars coming from the opposite direction – both in the intersection and the adjoining road. If such a car exists, and the cars are close enough, and the traffic lights do not override the give-way relationship, then a `giveWayToOnc` concept-instance is created.

Upon a give-way concept-instance creation, a `create` message is sent to the `deadlock` concept-frame, together with the identities of the two cars involved. This concept-frame checks for cycles in give-way chains (for instance, if car A gives way to car B, and car B gives way to car C, and car C gives way to car A, then nobody can move), and if such a cycle is found reports a legal deadlock or traffic jam.

The high-level interpretation of the network as a whole is generated by `deadlock` and the `give-way` concept-frames and sent to a special output file.

6.4 Trial Runs using XFIG

Initially, the system was run on diagrams of intersections generated by XFIG, a Unix drawing tool. This involved no low-level vision, just reading the XFIG data file to find the road, intersection and car token coordinates. After this point, the system ran in the same way as the full image system described below. Figure 6.4 shows a simple situation in an intersection.

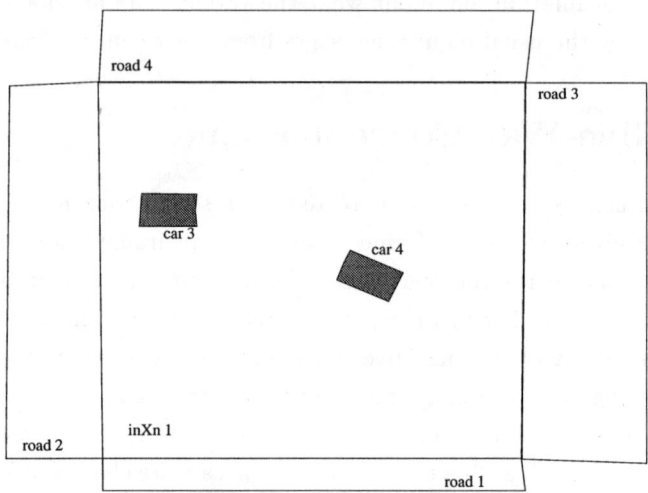

Figure 6.4: *Diagram showing a schematic intersection used for testing SOO-PIN. Roads and intersections are shown as line drawn quadrilaterals, and cars as grey filled rectangles. Each object is labeled with an identifier used in the interpretation system. This scene shows a simple give-way situation.*

Shown below is output generated by selected concept-frames from the run based on Figure 6.4.

```
output from car:
        *created id(car,3)
        *created id(car,4)
        *check relation in from inXn 1
        *check relation in from road 3 not found
        *inquiry from id(road,3) re relation in
        *check relation in from road 1 not found
        *check relation in from road 2 not found
        *check relation in from road 4 not found
        *update id(car,3) with [reln(in,id(inXn,1))]
        *update id(car,4) with [reln(in,id(inXn,1))]
        *inquiry from id(road,1) re relation in
        *inquiry from id(road,2) re relation in
        *inquiry from id(road,4) re relation in
```

After the two cars were created, together with the roads and intersection, they began by sending check messages exploring spatial and compositional relations to the next level concept-frames, which can be seen in the check messages above. Success is shown in the update messages where the two cars were found to be in the intersection. The action then moved on to higher concept-frames, of which the output from right is shown below.

```
output from right:
        *inquiry regarding reln(composedOf,id(inXn,1))
        *inquiry regarding desc(from(south))
        *created id(right,4472)
        *inquiry to id(right,4472) re desc(from(any))
        *inquiry regarding reln(composedOf,id(inXn,1))
        *inquiry regarding desc(from(east))
        *inquiry to id(right,4472) re reln(composedOf,id(car,any))
        *update id(right,4472) with [reln(partOf,id(giveWayToOnc,4473))]
        *inquiry to id(right,4472) re reln(composedOf,id(inXn,any))
        *inquiry regarding desc(from(north))
        *inquiry to id(right,4472) re desc(from(any))
        *inquiry to id(right,4472) re reln(composedOf,id(inXn,any))
        *inquiry to id(right,4472) re desc(from(any))
        *inquiry to id(right,4472) re reln(composedOf,id(car,any))
        *inquiry regarding reln(composedOf,id(inXn,1))
```

After a couple of exploratory inquiries, one concept-instance of a right-turner was created, as was a straight. These two objects caused the various give-way concept-frames to begin generating inquiry messages, of which some can be seen above. giveWayToOnc was successful, as can be seen from the update message above.

```
output from giveWayToOnc:
        *check relation composedOf from id(straight,4470)
        *adding inst 4473 composed of id(straight,4470)
                id(right,4472) id(car,3) id(car,4)
        *check relation composedOf from id(right,4472)
```

giveWayToOnc was able to establish from the existence of id(straight,4470) that there was a right-turner coming from the opposite direction, and which was in range of it. Thus it created a new concept-instance, and generated the following interpretation:

```
*Give Way to oncoming:   id(car,4) turning right from east
        gives way to id(car,3) from west
```

Thus, briefly, it can be seen how the system was able to deduce the story behind the picture by a process of independent agents each working on their own concepts.

Below some more complex examples of SOO-PIN operating on XFIG diagrams of intersections are given, but because of the higher cardinality, it is too space-consuming to show all the messages that move around the system, so just the final interpretations are shown.

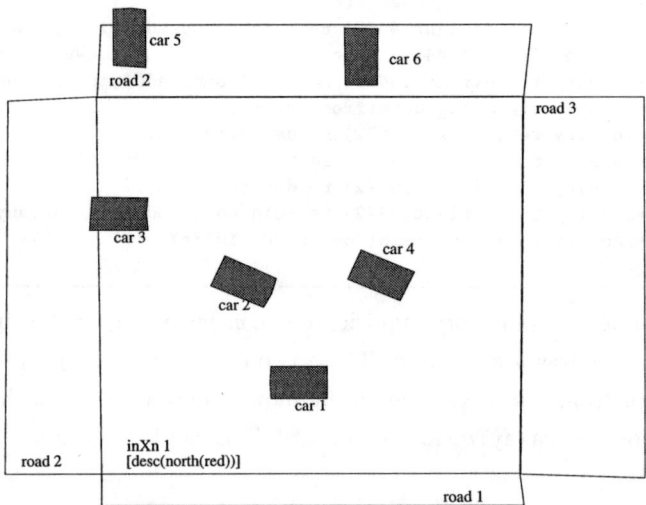

Figure 6.5: *Diagram showing a schematic intersection used for testing SOO-PIN. Roads and intersections are shown as line drawn quadrilaterals, and cars as grey filled rectangles. Each object is labeled with an identifier used in the interpretation system, the intersection is also labeled with the traffic light status. This scene shows a give-way situation modified by traffic lights.*

The output from Figure 6.5 is shown below:

```
*Give Way to oncoming:  id(car,2) turning right from west
       gives way to id(car,1) from east
*Give Way to oncoming:  id(car,4) turning right from east
       gives way to id(car,3) from west
*Give Way to Right overridden by Traffic Sign red :  id(car,6)
       from north gives way to id(car,4) from east
```

It can be seen how the two right-turners have given way to the two straight-throughers, and how the `giveWayToRt` concept-frame checked with the traffic light status of the intersection to modify its report on the give-way situation between `id(car,6)` and `id(car,4)`.

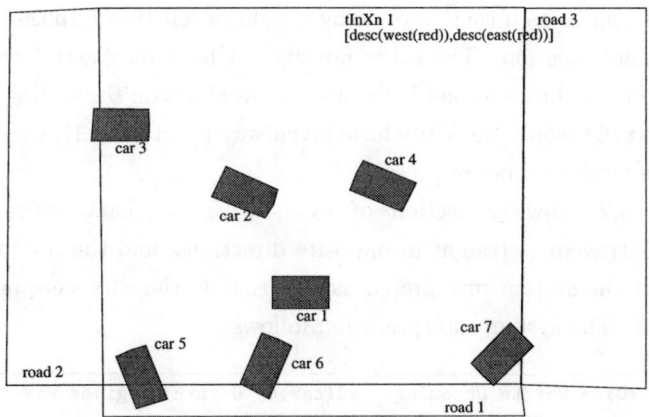

Figure 6.6: *Diagram showing a schematic T-intersection used for testing SOO-PIN. Roads and intersections are shown as line drawn quadrilaterals, and cars as grey filled rectangles. The through road is east-west. Each object is labeled with an identifier used in the interpretation system, the intersection is also labeled with the traffic light status. This scene shows a rather complex and unlikely configuration intended to demonstrate the system.*

Figure 6.6 shows a rather complex T-intersection in order to test the network with many objects and interactions, and also shows the slightly modified road rules that pertain in a T-intersection. The output is shown below:

```
*Illegal:  car id(car,4) from east is on the wrong side
       of T-intersection id(tInXn,1)
*Give Way to Left (T-inXn) overridden by Traffic Sign red :
       id(car,3) from west gives way to id(car,6) from south
*Give Way to oncoming:  id(car,2) turning right from west
       gives way to id(car,1) from east
*Give Way overridden by Traffic Sign red :  id(car,1) from
       east gives way to id(car,5) from south
*Give Way to Right overridden by Traffic Sign red :
       id(car,1) from east gives way to id(car,6) from south
*Give Way to left-turner:  id(car,2) turning right
       from west gives way to id(car,7) from east
```

The first novelty is that T-intersections generate an "illegal" report, as cars should not appear as if they were doing a right or left turn into the blank side of such an intersection. The other novelty (at least on Australian roads) is that cars on the through road have priority over cars on the ending road, and thus id(car,6) would normally have given way to id(car,3), except for the case where traffic lights are present.

Figure 6.7 shows a section of road, with two cars id(car,7) and id(car,8) traveling straight in opposite directions, and the rest of the cars doing what the system interpreted as U-turns, ie, the cars were not parallel to the road. The system interpretation follows:

```
*Give Way to Oncoming:  id(car,6) U-turning gives way
       to id(car,8) going straight
*Give Way to Oncoming:  id(car,1) U-turning gives way
       to id(car,8) going straight
*Give Way to Oncoming:  id(car,1) U-turning gives way
       to id(car,7) going straight
*Give Way to Oncoming:  id(car,14) U-turning gives way
       to id(car,7) going straight
*Give Way to Oncoming:  id(car,14) U-turning gives way
       to id(car,8) going straight
*Give Way to Oncoming:  id(car,2) U-turning gives way
       to id(car,8) going straight
*Give Way to Oncoming:  id(car,2) U-turning gives way
       to id(car,7) going straight
*Give Way to Oncoming:  id(car,3) U-turning gives way
       to id(car,7) going straight
```

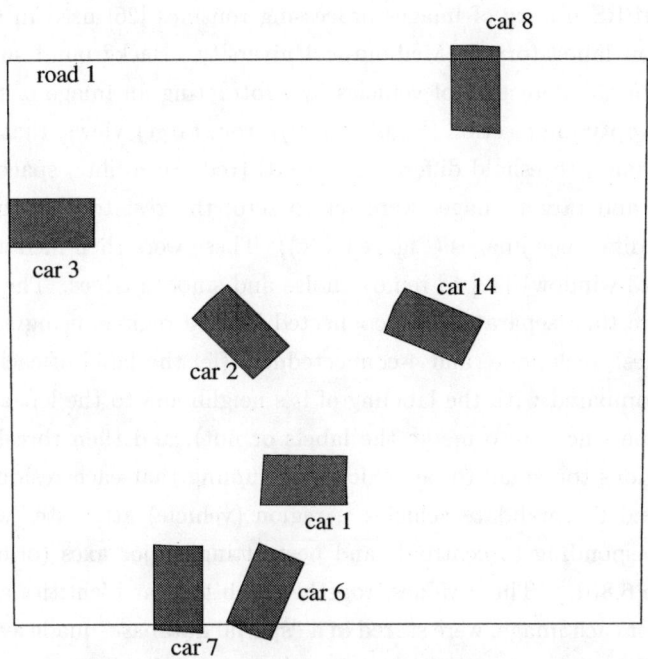

Figure 6.7: *Diagram showing a section of road used for testing SOO-PIN. The road is shown as a line drawn quadrilateral oriented north-south, and cars as grey filled rectangles. Each object is labeled with an identifier used in the interpretation system. This scene shows a rather complex configuration of cars in various orientations over a road.*

6.5   Low-Level Processing

For the actual intersection data, the system ran on images recorded on video-tape from a camera mounted above city intersections and the images were digitized using an Abekas Digital Video system (in 720x576, 24 bit RGB format), see Figure 6.8(a). The subsequent low-level processing was performed using the IPRS library of images processing routines [26] used in the Computer Vision laboratory at Melbourne University. Background subtraction was used for the detection of vehicles by subtracting an image of the corresponding empty intersection (Figure 6.8(b)) from target views, that is, pixels with less than a threshold difference in RGB (red-green-blue) space between the empty and target images were set to zero, the rest to one, thus creating binary difference images (Figure 6.8(c)). These were then median filtered (using a 3x3 window) [32] to remove noise and smooth edges. The resulting regions were then separated into connected labeled regions using an "equivalence tables" technique, and 4-connectedness (ie, the label of each pixel in turn was compared with the labeling of it's neighbours to the left and below to determine whether to merge the labels or not), and then thresholded to remove regions too small to be vehicles. Assuming that each residual region corresponded to candidate vehicles [1], region (vehicle) attributes were computed corresponding to centroids and best-fitting major axes (orientations) (see Figure 6.8(d)). These values, together with the car identities and frame numbers, for each image, were stored in a "spatial database" made available to SOO-PIN. The (constant) intersection and road coordinates were also stored in this database.

Note how in the example given in Figure 6.8, the dark car toward the top right is lost before final recognition. This is inherent in the technique used, a more sophisticated low level processing technique could have been used to pick up these cases, but for the purpose of this study the current technique was sufficient for "proof-of-concept" of the system.

---

[1] As will be seen, one benefit of the symbolic approach is that inappropriate segments can be checked for their consistency, updated or deleted by the symbolic analyses

Another inadequacy of the technique used is that it required manual se-
lection of the empty intersection, which, in a real world system, would have
to be done every few minutes to cope with changing lighting conditions. A
technique that could have been used to generate empty intersection images
automatically is to use a form of median filtering over time, ie, from continuous video of the scene, save 5 images from the last, say, 10 minutes (long
enough for the traffic lights to cycle twice). Then for each output image pixel
choose the median of the corresponding pixels in the 5 saved images. Since
most parts of the scene are free of cars most of the time, this yields an image
of an empty intersection.

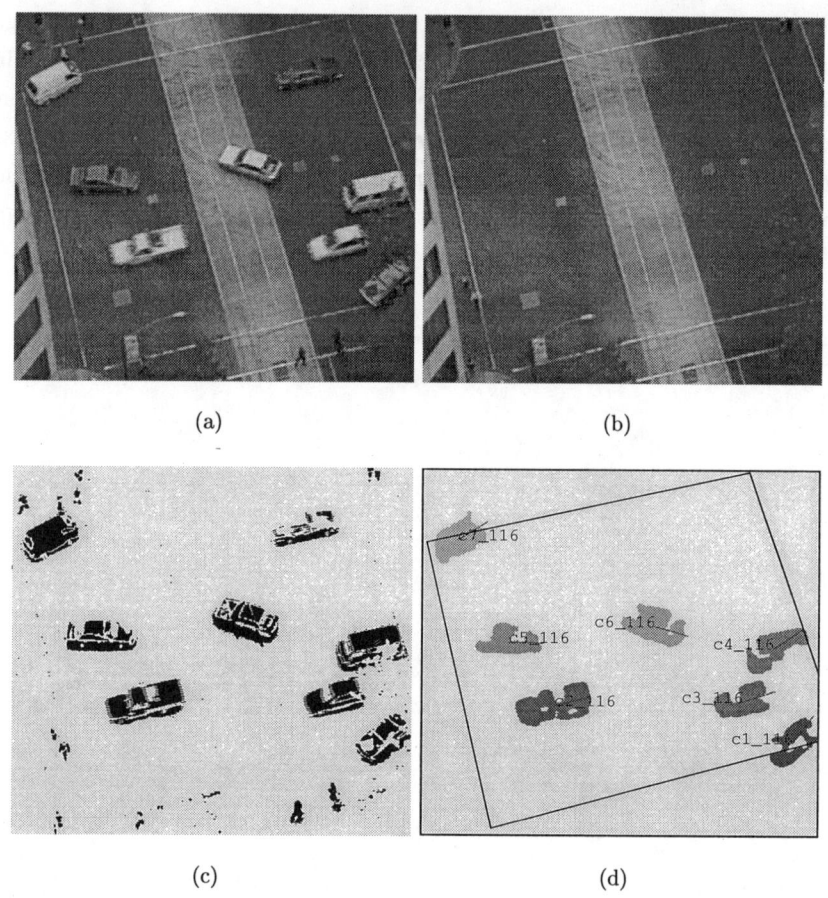

Figure 6.8: *Processing steps of real image: (a) Original image of intersection (b) Empty intersection (c) Raw differenced binary image (d) Median filtered, size filtered and labeled image, the lines on the cars are the major axes, the skew rectangle is the intersection boundary (determined manually).*

6.6 Trial Runs on Real Images

Using the image shown in Figure 6.8 as input into the traffic network described above yields the following interpretation:

```
*Give Way to oncoming:  id(car,c5_116) turning right from west gives
        way to id(car,c3_116) from east.

*Give Way to left-turner:  id(car,c5_116) turning right from west
        gives way to id(car,c4_116) from east.

*Give Way to oncoming:  id(car,c5_116) turning right from west gives
        way to id(car,c2_116) from east.

*Give Way to left-turner:  id(car,c5_116) turning right from west
        gives way to id(car,c1_116) from east.

*Give Way to left-turner:  id(car,c6_116) turning right from east
        gives way to id(car,c7_116) from west.
```

It can be seen how the system has interpreted id(car,c4_116) as a left turner, when clearly it was going straight ahead. The trouble was that the car was partly obscured by a tree, and thus the major axis of the car was skewed. This resulted in the left turn interpretation. Later in this study, this problem is rectified by the detection of car velocity.

The next example used an image from another intersection (see Figure 6.9), taken from a much lower camera resulting in an oblique view. This image demonstrated the usefulness of normalizing the image with an affine transformation onto a square. Potential problems with such normalization are that, firstly, it will not work if the images of the cars overlap due to perspective. Secondly, it has the effect of moving the centroid of the cars, when projected onto the plane of the intersection, further away from the camera. Both of these effects limit the amount of obliqueness, and thus the minimum height of the camera, that is acceptable.

The interpretation generated from Figure 6.9 is as follows:

```
*Give Way to left-turner:  id(car,c5_150) turning right from west
        gives way to id(car,c1_150) from east.
```

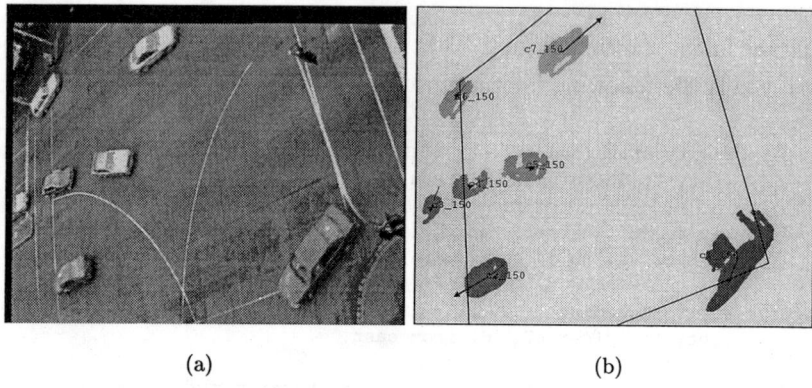

(a)   (b)

Figure 6.9: *(a) Original image of intersection (b) Median filtered, size filtered and labeled image, the lines on the cars are the major axes, the skew rectangle is the intersection boundary (input manually).*

This is correct, as judged by eye, but it is not the full story. It can be seen that car id(car,c4_150) was also turning right, but because its major axis was roughly aligned with the road, it was labeled as a car traveling straight ahead.

6.7   Traffic Trial Summary

In this chapter we have demonstrated the SOO-PIN concept implemented in a traffic scenario. This has been a useful exercise as this involves quite a complex network of frames to implement. Initially, it was implemented to run from XFIG diagrams, which demonstrated the capability of the system over a broad range of input situations. It was then shown how the system runs on real images, which, while more complex to process initially, results in simpler, more restricted traffic situations (in general, when videotaping traffic, one very seldom, if ever, sees an interesting traffic law violation). It was found that the system performed quite satisfactorily on these images, producing interpretations that the human observer would make.

Some limitations of the system resulted from using the orientation (major axis) of the cars, with their position in the intersection, to judge their heading and activity. This resulted in some inaccuracy in judging the direction the car was heading with consequent misinterpretation of the car's activity. This could be partly rectified if the velocity of the vehicles could be determined, which will be dealt with in a later chapter.

Another potential problem is the number of context dependent numerical thresholds involved in the implementation, for instance, the allowable angle that a car can be offset from straight ahead before it is judged to be doing another activity. This calls for some kind of measure of confidence in the judgments made within the various concept-frames to be passed around the network in the messages. This is dealt with in the next chapter.

Chapter 7

Uncertainty

7.1 Introduction

In vision systems, uncertainty comes from a number of sources. For instance, low-level object detection can return belief values deriving from identifying a given segment as a certain object, spatial predicate instantiation can return uncertainty from mapping continuous variables onto symbols, and high-level systems can involve uncertainty in deriving one proposition from others through such processes as induction. In SOO-PIN, there is a need to cope with uncertainty, at least to the extent of tracking it from predicate to predicate so that the final interpretations reflect the degree of uncertainty involved with their derivations.

There are two broad approaches to using uncertainty calculi in networks, the Bayesian approach and the Dempster-Shafer (DS) approach. The first is described by Pearl [52]. He shows how a network of nodes (in this case Markov trees) propagates belief. In this approach, each node represents a variable whose possible values are assigned probabilities (a frame of discernment), connected by links with the conditional probabilities expressed as a matrix. He distinguishes evidential support from causal support, which, in general, convey belief in opposite directions. In acyclic networks it is important to make that distinction as it prevents a change in probability at one node feeding back on itself and giving itself credence. One application of this approach is that of Huang et al[41] who use a Bayesian belief network and inference engine (HUGIN[3]) in sequences of highway traffic scenes to produce high-level concepts like "lane_change" and "stalled". A tool using Bayesian networks based on Pearl's ideas is BaRT [40] which is used to classify ship images.

Bogler [11] has pointed out that in the case of data fusion and object recognition, the Bayesian approach has limitations, namely:

- sensors provide information at varying levels of abstraction, and, for instance, if a sensor says "the target is of type A, B or C" (where A, B and C are possible targets), with a certain probability, the Bayesian approach forces the system to divide the probability equally between the three types.

- the sensors are required to have a complete and accurate knowledge of both the prior probability distribution and the conditional probability matrices. If they do not have this data, the Bayesian approach forces them to guess.

- if sensors give contradictory readings, the Bayesian approach has no formal means of dealing with this. The usual solution is to distribute various likelihoods where they are unknown.

Bogler shows how the Dempster-Shafer approach solves these difficulties, and how this approach is used for target recognition.

Lowrance et al [44] in a report on evidential reasoning systems for the US navy, like Pearl, deals with Markov trees that convey belief bidirectionally, in this case from a Dempster-Shafer perspective (DS is explained further in Section 7.2 below). In Lowrance's system, each node is associated with a "frame of discernment" where belief is expressed in the form of a mass distribution over the set of subsets of the frame. Links are in the form of "compatibility mappings" between the sets of subsets of the respective frames of discernment, and thus play the role of Pearl's conditional probability matrices. A pair of nodes will have a compatibility mapping in each direction, and they are, like Pearl, careful to avoid positive feedback (which would result in propositions supporting themselves). A number of systems have been developed using the

Lowrance approach. For instance, Garvey [31] describes a helicopter route planning system in which each pixel of a topographic map is attached to a network of nodes representing such things as vegetation cover, terrain type, visibility and overall danger.

Wesley [67] presents a system for labeling regions of 2D monochromatic scenes using Dempster-Shafer evidential reasoning – as described by Lowrance. He limits the frames of discernment to the interpretation of given regions, and discusses independence of evidence, noting the differences between the DS and Bayesian interpretation. Wesley argues that it is necessary to develop a formal model to account for dependencies between knowledge sources, – a poignant issue which he does not pursue.

Baldwin [6] describes a system (FRIL) which uses the basic Dempster-Shafer formalism in a logic-programming inference system. This system is to be distinguished from the approaches of Pearl and Lowrance in that it does not deal with combining evidence within a frame of discernment, but rather with combining belief between independent propositions. Later (Section 7.4.1) we will show how this approach is particularly useful for vision.

The Dempster-Shafer uncertainty calculus is discussed below in more detail, along with some verifications of Baldwin's results which were not shown in his original paper [6]. Following this, some criticisms and caveats regarding Baldwin are dealt with, and the chapter finishes by showing how uncertainty is implemented into the SOO-PIN system.

7.2   Dempster-Shafer Theory

Like Bayesian uncertainty calculus, Dempster-Shafer uncertainty calculus starts from a set of exhaustive and mutually exclusive propositions in a "frame of discernment". Again, as in Bayes, there is an assignment of weights, but in this case the weight is assigned to *sets* of these propositions, including the set of all the propositions in the frame of discernment. This formulation allows for an expression of degrees of ignorance. For instance, if the frame of discernment is a set of possible burglars, then if the burglar is found to be male, then a weight can be associated with the set of all male burglars. In Bayes, without prior knowledge, one would have to give an equal increment of weight to each *individual* male burglar, thus implicitly providing more information than we actually have. This weight assignment is called a "mass distribution", and obeys the following:

$$m : 2^\Theta \mapsto [0, 1] \qquad (7.1)$$

$$\sum_{A_i \subseteq \Theta} m(A_i) = 1$$

$$m(\emptyset) = 0$$

where Θ is the frame of discernment, and 2^Θ is the set of all subsets of Θ.

Uncertainty of a proposition P is interpreted via its "support" and "plausibility". The support is given by the sum of all the weights attached to subsets of the proposition, ie:

$$S(P) = \sum_{A \subseteq P} m(A) \qquad (7.2)$$

The plausibility of a proposition P is the difference between certainty and the support of $\neg P$, ie:

$$P(P) = 1 - S(\neg P) \qquad (7.3)$$

The support and plausibility are referred to below as the "belief pair" or "belief interval" of a proposition.

Since the sum of all weights is unity, then $S(P) + P(P) \le 1$ (Lowrance et al [44, p7]). If in a degenerate case each and only singleton propositions are given weight by the mass assignment, then the belief pair $[S, P]$ has the one value $S = P$ and is equal to the Bayes probability. In this way Dempster-Shafer uncertainty can be viewed as a superset of Bayesian probability. Pearl, however, views DS as philosophically quite distinct to Bayes, namely, that the DS belief values deal with the probability of the *provability* of propositions, whereas Bayes deals with the probability of the *truth* of them. In Pearl's view, a model for DS is provided by thinking of each frame of discernment as having a timer that assigns "truth" to each proposition for a fraction of the time corresponding to the value of the mass of that proposition. One then finds the logical outcome of the assignments for each time instance, and ascribes mass to the outcomes by the proportion of the time they are true. This model produces the required DS calculus as described below.

7.2.1 Combining Evidence within a Frame of Discernment

If we have evidence for a proposition P in the form of a mass distribution m_1 and other independent evidence in the form m_2, then Dempster's rule of combination [24] allows us to calculate the combined mass distribution m_3 reflecting the combined evidence for P:

$$
\begin{aligned}
m_3(P) &= m_1 \oplus m_2(P) \qquad\qquad (7.4)\\
&= \frac{1}{1-\kappa} \sum_{A_i \cap A_j = P} m_1(A_i)m_2(A_j)\\
\kappa &= \sum_{A_i \cap A_j = 0} m_1(A_i)m_2(A_j)
\end{aligned}
$$

The factor κ is referred to as the *conflict* between the evidence represented by m_1 and m_2 (Lowrance et al [44, p8]), ie, if there is conflict, there are pairs of sets with mass assigned by m_1 and m_2 respectively with no intersection.

This rule allows us to find the combined weight of evidence for an hypothesis from various sources. For instance, if one source of evidence suggests that a burglar is male with a mass distribution giving weight to the subset of males (of, say, 0.8), and another source gives weight to the subset of redheads (of, say, 0.7), then we can use Dempster's rule to calculate the combined mass and possibly suggest the culprit. Assuming the males are Alex, Bill and Chris, and the redheads are Chris and Diane, then using Equation 7.4, we need to calculate m_3 given the two distributions from the evidence, namely m_1 giving weight 0.8 to males and 0.2 to Φ (all the suspects), and m_2 giving weight 0.7 to redheads and 0.3 to Φ. From Table 7.1 it can be seen that the only non-empty intersecting subsets of the two distributions are *Chris, male, redhead*

$\cap$	male	Φ
redhead	Chris	redhead
Φ	male	Φ

Table 7.1: *Evidence combinations for the burglar example. This table shows the non-empty intersections of subsets assigned non-zero weight by each evidence source.* Φ *represents all the suspects,* male *is the set of males, and* redhead *is the set of redheads.*

and Φ. There are no empty intersecting sets, therefore $\kappa = 0$ and $\frac{1}{1-\kappa} = 1$. The calculation of m_3 is as follows:

$$m_3(Chris) = \frac{1}{1-\kappa} * m_1(male) * m_2(redhead) = 0.8 * 0.7 = 0.56$$

$$m_3(redhead) = \frac{1}{1-\kappa} * m_1(\Phi) * m_2(redhead) = 0.2 * 0.7 = 0.14$$

$$m_3(male) = \frac{1}{1-\kappa} * m_1(male) * m_2(\Phi) = 0.8 * 0.3 = 0.24$$

$$m_3(\Phi) = \frac{1}{1-\kappa} * m_1(\Phi) * m_2(\Phi) = 0.2 * 0.3 = 0.06$$

$$(7.5)$$

It can be seen that the highest belief in the combined evidence is for Chris as the culprit, with a weight of 0.56.

7.2.2 Combining Independent Propositions

Dempster-Shafer works well in situations where the frame of discernment is clear, with evidence bearing upon this frame of discernment from independent sources. However, as explained in Section 7.4.1, in Computer Vision, frames of discernment are quite simple, but evidence bears upon independent propositions. Therefore it is necessary to consider Baldwin's [6] work, where he deals with belief assigned to independent propositions. In this section a result shown by Baldwin in which he combines belief intervals of two propositions in independent frames of discernment is justified. Another reason for exploring this is that not only is his approach compatible with vision, but such combinations fit nicely within logic programming. For instance, a typical Horn clause might be $P \Leftarrow A \wedge B$ (P is proven if both A and B are). To include uncertainty one can use Baldwin's combination of independent propositions rules to compute the uncertainty of P given the uncertainty A and B.

Given two propositions A and B with belief intervals $[\mathcal{S}(A), \mathcal{P}(A)]$ and $[\mathcal{S}(B), \mathcal{P}(B)]$, then Baldwin's combination rules are, firstly, conjunction:

$$\mathcal{S}(A \cap B) \;=\; \mathcal{S}(A).\mathcal{S}(B) \tag{7.6}$$

$$\mathcal{P}(A \cap B) \;=\; \mathcal{P}(A).\mathcal{P}(B) \tag{7.7}$$

secondly, disjunction:

$$\mathcal{S}(A \cup B) \;=\; \mathcal{S}(A) + \mathcal{S}(B) - \mathcal{S}(A).\mathcal{S}(B) \tag{7.8}$$

$$\mathcal{P}(A \cup B) \;=\; \mathcal{P}(A) + \mathcal{P}(B) - \mathcal{P}(A).\mathcal{P}(B) \tag{7.9}$$

As can be seen, these rules are very similar to the equivalent Bayesian rules. These rules allow propagation of uncertainty values through a network, each node passing a belief pair in a message to other nodes which can calculate the belief pairs for various logical statements. Of course, as explained in Section 7.3, it is important to ensure that the network contains no dependency loops, since this breaks the precondition for the combination rules.

7.2.2.1   Conjunction Rule

In his paper [6], Baldwin did not derive in detail the conjunction in the combination rule Equation 7.6. This rule is now justified.

If proposition B is in Frame of Discernment Φ and proposition A is in Frame of Discernment Θ, and we have $\mathcal{S}(B)$ and $\mathcal{S}(A)$, then we want to know $\mathcal{S}(A \cap B)$. In order to combine propositions in independent, "orthogonal"

Frames of Discernment we need to work in the cross product frame $\Theta \times \Phi$.

From the definition of the support of a proposition given in Equation 7.2,

$$S(B) = \sum_{X \subseteq B} m_2(X) \tag{7.10}$$

then

$$S(B) = m_2(X_1) + m_2(X_2) \cdots m_2(X_n)$$

where $m_2(X_i)$ are non-zero, and m_2 is the mass assignment for the frame Φ. Similarly for A:

$$S(A) = m_1(Y_1) + m_1(Y_2) \cdots m_1(Y_m)$$

From the same definition,

$$S(A \cap B) = \sum_{X \subseteq A \cap B} m_3(X)$$

$$= \sum_{X \cap Y \subseteq A \cap B} m_3(X \cap Y)$$

Since the only mass carrying members of $\Theta \times \Phi$ are sets of the form $X \times \Phi \cap Y \times \Theta$ where $X \subseteq \Theta$ and $Y \subseteq \Phi$,

$$S(A \cap B) = \sum_{i,j} m_3(X_i \times \Phi \cap Y_j \times \Theta)$$

from the enumerated non-zero mass subsets of Θ and Φ given above.

Now from the definition of combined mass assignments given in Equation 7.4,

$$m_3(X_i \times \Phi \cap Y_j \times \Theta) = \frac{1}{1 - \kappa} \sum_{V \times \Phi \cap W \times \Theta = X_i \times \Phi \cap Y_j \times \Theta} m_1(V).m_2(W)$$

$$= m_1(X_i).m_2(Y_j)$$

as there is only one pair $V \times \Phi \cap W \times \Theta$ equal to $X_i \times \Phi \cap Y_j \times \Theta$, namely themselves. The κ is 0 as

$$\kappa = \sum_{V \times \Phi \cap W \times \Theta = \phi} m_1(V).m_2(W) \tag{7.11}$$

and V and W are orthogonal and always have non-zero intersection.

Thus

$$S(A \cap B) = \sum_{i,j} m_1(X_i).m_2(Y_j)$$

$$= (\sum_i m_1(X_i)).(\sum_j m_2(Y_j))$$

$$= S(A).S(B) \tag{7.12}$$

again from the definition of the Belief function.

7.2.2.2   Disjunction Rule

For disjunction (Equation 7.8) we can assume without loss of generality that the mass distribution in Φ is limited to $m_2(B), m_2(\neg B)$ and $m_2(\Phi)$ as the mass of all other subsets can be subsumed under the total frame Φ, similarly for Θ. This means $S(B) = m_2(B)$, $S(\neg B) = m_2(\neg B)$ and $S(\Phi) = m_2(\Phi) = 1 - m_2(B) - m_2(\neg B)$ (this latter because mass distributions must sum to 1). Thus support for B is $m_2(B)$, plausibility of B is $1 - m_2(\neg B)$, and doubt about B is $1 - m_2(B) - m_2(\neg B)$.

To deal with the disjunction $A \cup B$ it will help to use the following Table 7.2. The masses assigned to the sets in the table are calculated from Equation 7.12. These values follow clearly from the conjunction rule above, except for the cases involving Θ or Φ. An example is shown below:

$$S(A \cap \Phi) = \sum_{X \subseteq A \cap \Phi} m_3(X)$$

which since the only non-zero mass subset is itself

$$= m_3(A \cap \Phi)$$

$\Theta \times \Phi$	B	$\neg B$	Φ
A	$A \cap B$ $m_1(A).m_2(B)$	$A \cap \neg B$ $m_1(A).m_2(\neg B)$	$A \cap \Phi$ $m_1(A).m_2(\Phi)$
$\neg A$	$\neg A \cap B$ $m_1(\neg A).m_2(B)$	$\neg A \cap \neg B$ $m_1(\neg A).m_2(\neg B)$	$\neg A \cap \Phi$ $m_1(\neg A).m_2(\Phi)$
Θ	$\Theta \cap B$ $m_1(\Theta).m_2(B)$	$\Theta \cap \neg B$ $m_1(\Theta).m_2(\neg B)$	$\Theta \cap \Phi$ $m_1(\Theta).m_2(\Phi)$

Table 7.2: *Masses assigned to the members of $\Theta \times \Phi$ with non-zero mass*

By the definition of combined mass assignments given above,

$$m_3(A \cap \Phi) = \frac{1}{1 - \kappa} \sum_{V \cap W = A \cap \Phi} m_1(V).m_2(W)$$

Since, by the assumption of limited mass distribution, the only such sets V and W are A and Φ, and the factor κ, shown in Equation 7.11, is 0 (for the same reason as in Equation 7.11), then

$$\mathcal{S}(A \cap \Phi) = m_1(A).m_2(\Phi)$$

$$= m_1(A).(1 - m_2(B) - m_2(\neg B))$$

By a similar argument,

$$\mathcal{S}(\Theta \cap B) = m_2(B).(1 - m_1(A) - m_1(\neg A))$$

To find $\mathcal{S}(A \cup B)$ use the support definition Equation 7.10 above:

$$\mathcal{S}(A \cup B) = \sum_{X \subseteq A \cup B} m_3(X)$$

$$= m_3(A \cap B) + m_3(A \cap \neg B) + m_3(\neg A \cap B) + m_3(A \cap \Phi) + m_3(\Theta \cap B)$$

which from Equation 7.12 and Table 7.2 is:

$$
\begin{aligned}
&= & m_1(A).m_2(B) + m_1(A).m_2(\neg B) + m_1(\neg A).m_2(B) + \\
& & m_1(A).(1 - m_2(B) - m_2(\neg B)) + (1 - m_1(A) - m_1(\neg A)).m_2(B) \\
&= & m_1(A) + m_2(B) - m_1(A).m_2(B) \\
&= & \mathcal{S}(A) + \mathcal{S}(B) - \mathcal{S}(A).\mathcal{S}(B)
\end{aligned}
$$

This is the required result given in Equation 7.8.

7.2.2.3   Plausibility

The plausibilities for conjunction and disjunction in Equations 7.6 and 7.8 follow from Equation 7.3 and the combination rules for support.

Firstly the case of conjunction:

$$
\begin{aligned}
\mathcal{P}(A \cap B) &= 1 - \mathcal{S}(\neg(A \cap B)) \\
&= 1 - \mathcal{S}(\neg A \cup \neg B) \\
&= 1 - \mathcal{S}(\neg A) - \mathcal{S}(\neg B) + \mathcal{S}(\neg A).\mathcal{S}(\neg B) \\
&= (1 - \mathcal{S}(\neg A)).(1 - \mathcal{S}(\neg B)) \\
&= \mathcal{P}(A).\mathcal{P}(B)
\end{aligned}
\tag{7.13}
$$

Secondly the case of disjunction:

$$
\begin{aligned}
\mathcal{P}(A \cup B) &= 1 - \mathcal{S}(\neg(A \cup B)) \\
&= 1 - \mathcal{S}(\neg A \cap \neg B) \\
&= 1 - \mathcal{S}(\neg A).\mathcal{S}(\neg B) \\
&= 1 - (1 - \mathcal{P}(A)).(1 - \mathcal{P}(B)) \\
&= \mathcal{P}(A) + \mathcal{P}(B) - \mathcal{P}(A).\mathcal{P}(B)
\end{aligned}
\tag{7.14}
$$

7.2.2.4 Combining Belief from Two of N Events

We are now in a position to derive an algorithm `belTwoOfN` to calculate the belief for the case where an object exists granted the existence of two components and a relationship between them out of a population of many possible components. For instance, if a cube D exists when any two of sides A, B or C, and the edge between them, exist, then:

$$D \Leftarrow (A \wedge B) \vee (B \wedge C) \vee (C \wedge A) \tag{7.15}$$

In this case, the usual Baldwin formulae (Equation 7.6, etc.) cannot be used, as the disjunctions are not independent, nor is there any transformation that turns them into a set of independent expressions. To calculate this type of expression, a more fundamental treatment is needed.

More generally, a compound object exists given the existence of any two of its components (and a relationship between them), and given the belief in all its components (and their relationships), then the total belief in the object is calculated as follows:

If there are M components then there are $N = M.(M-1)/2$ potential (symmetric) relationships. The belief can be calculated by analogy with the probability calculation given by Feller[29, p98–109], where the formula for the probability of at least one event occurring among N is:

$$P = S_1 - S_2 + S_3 - S_4 + ...(-1)^{(N-1)} S_N. \tag{7.16}$$

where $S_1 = \sum P_i$, $S_2 = \sum P_{ij}$, $S_3 = \sum P_{ijk} \cdots$, where, for instance, $P_{ijk} = P\{R_i R_j R_k\}$ is the probability of $R_i \wedge R_j \wedge R_k$, and R_i is the ith relationship (event) in the list of N. This formula is quite general as there is no assumption about the independence of the R_i within the conjunction.

To move this into the Dempster-Shafer perspective it is necessary to deal with support and plausibility separately. In deriving his formula, Feller[29] computes probabilities of events, say A, from the frequency of sample points being "contained" in it: a Venn diagram model. The Venn diagram model can be used for support and plausibility by saying A occurs when a sample point is in it (for support) and a sample point is **not** in its complement (for plausibility). Note that this DS model assumes that a frame of discernment consists only of a proposition and its negation. With this alteration, Feller's argument can be used for both support and plausibility, and thus the algorithm below is identical for both. This symmetry between support and plausibility is also visible in Equations 7.6 to 7.9.

In the following an algorithm is developed to find the belief in an object given its parts and their relationships (in the form of a list of relationships, each relationship containing its belief and its endpoints with their belief). Note that only the description for support is given and, by symmetry, plausibility follows.

The algorithm is based on Equation 7.16. First, take the power set of the list of relationships giving a set of lists of all lengths from 1 to the length of the original list. Using Equation 7.16, each $S_n = \sum P_{ij\cdots q}$ is the sum of the supports for all conjunctions of relations of cardinality n. To calculate a specific $P_{ij\cdots q}$ first find the union of the endpoints of the n relationships $R_i, R_j, \cdots R_q$, say $O_u, u = 1 \cdots k$ where the union has k endpoints. Then

$$P_{ij\cdots q} = \mathcal{S}(R_i).\mathcal{S}(R_j)\cdots\mathcal{S}(R_q).\prod_{u=1}^{k}\mathcal{S}(O_u) \qquad (7.17)$$

where $\mathcal{S}(R_i)$ is the support of R_i. This expression follows from Equation 7.6 since it is assumed that the R_i and O_u are independent.

This is calculated for all such $P_{ij\cdots q}$ and summed to give S_n above. These S_n, for each n, are then summed, alternating signs in accordance with Equation 7.16, thus giving the total support for at least one relation existing among N possible. This algorithm is used extensively in SOO-PIN to calculate belief pairs.

This algorithm has been coded in Parlog++. To demonstrate it, say there are three objects A, B and C, and relations between them X(A,B), Y(B,C) and Z(C,A), and a compound object that exists, granted any pair of the objects (and their relations) exist. Then the belief in that compound is given by

belTwoOfN. Table 7.3 shows the input belief support values for these objects
and relations, together with the resulting support output by the algorithm
(plausibility is analogous):

The first group of three rows shows that when only two of the three objects

A	B	C	X(A,B)	Y(B,C)	Z(C,A)	belTwoOfN
1	1	0	1	1	1	1
1	0.5	0	1	1	1	0.5
1	0	0	1	1	1	0
1	1	0.5	1	1	1	1
1	0.5	0.5	1	1	1	0.75
0.5	0.5	0.5	1	1	1	0.5
1	1	1	0.5	1	1	1
1	1	1	0.5	0.5	1	1
1	1	1	0.5	0.5	0.5	0.875
0.5	0.5	0.5	0.5	0.5	0.5	0.296875

Table 7.3: *Example of the algorithm belTwoOfN in action, where the support for three
objects A, B and C, and relations between them X(A,B), Y(B,C) and Z(C,A) is given in
the rows, the output from the algorithm is shown on the right.*

exist (ie, support for C is zero), then belTwoOfN behaves like the conjunction
of A and B under Equations 7.6 (the relationships X, Y and Z have been set to
1 for simplicity). The next group of three rows demonstrate that calculating
the belief from A, B and C given that the object exists if any pair exists, is
not a straightforward conjunction or disjunction. The last triple shows that
the relationship between the X, Y and Z is simply that of disjunction (setting
the objects A, B and C to 1).

   This table shows that the algorithm belTwoOfN gives the right sort of
answers for N = 3. It has also been verified with a form of truth table for
N=3 and 4. Experiments with higher N give the intuitively expected results.

7.3   Problems with Dempster-Shafer

Pearl [52, p447] gives an example of the application of Dempster-Shafer that comes up with an unintuitive answer. Baldwin [6, p153] gives the same example and derives an acceptable answer. The well-known example in AI is:

> Tweety is a penguin, penguins do not fly, penguins are birds and birds usually fly. Does Tweety fly?

More formally, (where the ϵ are small compared to 1, and the support and plausibility for a clause follow it in square brackets):

$$
\begin{aligned}
fly(X) &\Leftarrow isPenguin(X) : [\epsilon_1, \epsilon_1] \\
fly(X) &\Leftarrow isBird(X) : [1 - \epsilon_2, 1 - \epsilon_3] \\
isBird(X) &\Leftarrow isPenguin(X) : [1, 1] \\
isPenguin(tweety) &: [1, 1]
\end{aligned}
$$

The query is: $fly(tweety)$ with what certainty? Applying Baldwin's combination rules (Equation 7.6) one gets support for $fly(tweety)$ of $\approx \epsilon_1/\kappa$ and plausibility $\approx 1 - (\epsilon_2/\kappa)$, where $\kappa \approx \epsilon_1 + \epsilon_2$. Pearl criticizes this result, pointing out that apparently the flying ability of Tweety depends on the proportion of non-flyers among birds, even though this is irrelevant given that we know she is a penguin. However, this result is also equivalent to Baldwin's, if we use in his argument $\epsilon_1 = 0$, $\epsilon_2 = 0.1$ and $\epsilon_3 = 0$ giving the intuitive correct answer that support and plausibility for Tweety flying is zero. In this case, Pearl's analysis is more rigorous!

The problem is that the data is not independent, and Dempster's rule, and thus Baldwin's rules (Equations 7.6 to 7.9) above, are not applicable. If one removes the dual dependency, knowing the fact that penguins are birds and birds fly (as Pearl does for his Bayesian analysis of this example), then one gets the intuitively correct answer.

This rings a warning bell for the application of DS to logic programming (LP): one must ensure independence among antecedents of propositions. In normal LP, it does not matter if there are dependencies, propositions are proved true if any path to them is proven, hence in languages like Prolog there is no mechanism to identify or remove dependency. This will have to be performed either mechanically or by the programmer when applying DS to LP.

7.4 SOO-PIN and Uncertainty

In order to incorporate uncertainty measures in SOO-PIN, the uncertainty calculus needs to be compatible with logic programming (LP). Pearl claims [52, p416] that the Dempster-Shafer formalism is syntactically compatible with LP, although he cautions against the dependency problem noted above. Lowrance et al [44, p46] also suggests that DS is compatible with LP. Baldwin describes a full Dempster-Shafer logic programming system in [6], and (as described above) the DS combination rules for Horn clauses.

Dubois and Prade[28], in a useful critique of Baldwin's work, point out a problem with Baldwin's Support Logic Programming, namely that the support pair for logical implication "$B \Rightarrow A \ [\mathcal{S}, \mathcal{P}]$" does not have a mathematical definition, and is not equivalent to its declarative interpretation "$\neg A \vee B \ [\mathcal{S}, \mathcal{P}]$".

Baldwin's approach to uncertainty has been adopted for SOO-PIN, but with belief pairs attached to *facts* (ie, there is a car at position x with belief $[\mathcal{S}, \mathcal{P}]$) but not to *clauses*. This avoids the problem pointed out by Dubois and Prade but still allows the use of Baldwin's combination rules. It also allows the use of the existing Parlog++ syntax. In adopting Baldwin's approach, it is of course necessary to be careful to avoid combining evidence which is not independent.

7.4.1 Belief and Vision

In Computer Vision, there is the question of the meaning of "frames of discernment" as there is a huge number of potential frames, corresponding to the set of all possible image regions, even the power set of the pixels in the image! The proposition "blob A is a motorcycle" is not independent of the proposition "blob B is a car" if the blobs intersect, but they are also not in the same frame of discernment as these consist of sets of exhaustive and mutually exclusive propositions. Therefore in vision we have simple frames of discernment consisting of, for instance, "blob A is a car" and "blob A is not a car"; but with potentially complex dependencies between frames of discernment. Consequently, this is another reason beyond that given in Section 7.3, to be careful about dependency when combining evidence.

Richer frames of discernment occur when dealing with higher-level concepts founded on given objects. For instance, given a car A, the kind of turn it is doing is only one of a finite list. It is this "closed world" that characterizes frames of discernment.

7.4.2   Implementation of Uncertainty in SOO-PIN

We start by describing the data structures and then describe the logic involved in determining the existence of objects. Finally, the belief messages that are passed around the network are analyzed.

7.4.2.1   Data Structures

A structure called "belief" is defined, bel/2, containing the support and plausibility of a proposition, for example, bel(0.34,0.89). The identity of objects, which were previously given by the structure id(Type,Number) (see Section 4.4), are expanded to include the belief in the existence of the object, ie id(Type,Number,Bel) where Bel is the belief pair. Relations or descriptions also have belief attached to them, ie, reln(RelnType,OtherId,Bel) and desc(DescType,Bel). Thus belief is attached to everything substantive that can be said within the system.

7.4.2.2   Existence Checking

In SOO-PIN, a new concept-instance is created when its antecedents are found. This occurs when a message is received from one of the antecedents by the concept-frame, and it sends messages to the other antecedents to determine if the existence criteria are satisfied. For instance, in the case of

$$\text{unicycle} \Leftarrow \text{wheel} \land \text{seat} \land$$
$$\text{above(seat,wheel)} \land$$
$$\text{joined(seat,wheel)}$$

where the message is sent by *wheel*, the *unicycle* would send a message to *seat* inquiring if there is a seat joined to and above the sending wheel. To introduce belief, the belief values from all four antecedents in this rule are combined. This is quite straightforward, as the belief combination rules (Equations 7.6 to 7.9) are commutative and associative, their order is irrelevant, and can be combined as one operation on a list of beliefs. If the resultant belief is over some threshold (which can vary from object to object), then an instance is created.

There can be any number of logical statements (clauses) mediating between the existence of an object instance and its components. There may simply be a list of necessary components (and their interrelationships), in which case the belief in the object instance is the multiple of the beliefs in the components (and that of the interrelationships). More commonly, the object instance exists granted any two (or more) from a list of potential components. In SOO-PIN, the belief is given by the algorithm shown in Section 7.2.2.4 above. In every case, the justification for an object instance's existence is given by applying a logical statement to a list of component relationships - the "justification list" which is held in the inst structure of each concept-frame. The belief in the object follows from the logical statement and the list, but in some cases the relationship between belief and the logical statement could be difficult to find. The algorithm belTwoOfN in Section 7.2.2.4 is an example of this.

7.4.2.3 Belief Updating

In this chapter we have dealt with adding a belief measure to SOO-PIN. However, so far we have not dealt with the fact that belief values can change over time. This section introduces an extension to SOO-PIN which allows belief updating via message passing between concurrent concept-frames. Such an extension is described below, but has not been implemented.

If a message from an antecedent object is received and found to be a new component of an existing instance, the belief in that compound instance may need to be changed. This involves recalculating its belief from all its components and the relevant relationships, which will be straightforward as these are all stored in the instance's justification list. If belief in the instance does change, then any other objects in the network that depend upon it will need to be informed. Therefore there will be a new message belUpd that is sent to all such objects with the senders new belief value. This message belUpd would replace a previous message negCheck that was used to convey knowledge of the deletion of instances to dependent objects. This role is taken by sending belUpd with the belief value of bel(0,0).

Upon receiving the message belUpd, an object will recalculate the appropriate instance's belief using the new belief in the message plus the belief in its other components from its justification list. If changed, it will then send new belUpd messages of its own. In this way, negative information becomes a natural part of the processing of SOO-PIN, the belief values propagate through the network until a new interpretation is made. Dependency loops can be avoided because such messages will always be sent higher in the abstraction space.

Note that with this extension, SOO-PIN has a similarity with Assumption-based Truth Maintenance Systems (ATMS) [22], namely, the justification for believing in an instance of an object is stored with the instance in the form of its justification list, and updating the belief in one concept results in automatic updates to the belief in dependent concepts.

7.4.2.4 Procedural Subroutines

SOO-PIN calls procedural subroutines written in the language C to calculate certain spatial predicates, for instance: reln(in,car,inXn), reln(east,-inXn,road), reln(target,car,car) and turns(right,car). The routine that returns the turns predicate also returns a belief pair based on the proximity of the car to the boundaries in the state space that determines what turn the car is doing. That is, if the car is very close to such a boundary, there is some evidence that the car is doing some other turn in the intersection (that of the other side of the state space boundary), and thus the system returns an uncertainty closer to bel(0.5,0.5) In Figure 7.1 there is a row of cars in an intersection that lies across the boundary separating right-turners from the east from right-turners from the west, the procedural routine turn.c returns the following:

```
turn car 1 inXn 1
    [car,1,inXn,1,east,right,0,bel(0.916,0.916)].
turn car 2 inXn 1
    [car,2,inXn,1,east,right,0,bel(0.712,0.712)].
turn car 3 inXn 1
    [car,3,inXn,1,west,right,1,bel(0.584,0.584)].
turn car 4 inXn 1
    [car,4,inXn,1,west,right,1,bel(0.903,0.903)].
```

It can be seen how as the border between right turn from the east and that from the west is approached, the belief pair returned drops from 0.916 to 0.712 before the returned activity switches. This belief approaches 0.5 as the border is approached within a "border zone" whose characteristic scale is set for each context, for instance, here it was set to 0.05, meaning 1/20th of the intersection side length, being about the size of a car in the intersection.

7.4.2.5 Belief Runtime Experiments

Uncertainty can come about through the input to SOO-PIN. For instance, the low-level vision system used to find cars in the traffic scenes can produce a "probability" measure p with each car. This is input to SOO-PIN with a belief of $bel(p, 1)$ which means the input object instance has support of p and no evidence to the contrary. Secondly, the procedural calls made by SOO-PIN to determine things like nearness of one object to another also return uncertainty (in this case behaving like a "fuzzy" measure, eg, car A is 0.45 near to car B). After the uncertainty values pass up through the system, it outputs descriptions like "car A is 70% likely to be giving way to car B". A

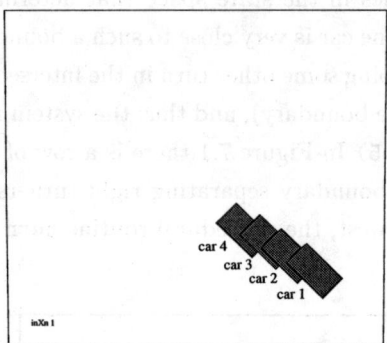

Figure 7.1: *Row of cars straddling the boundary between right turns from the east versus the west, demonstrating the drop in belief in the calculated turn activity as the boundary is approached.*

numerical output is useful if the output is to be passed to another computer system, but if it is to be read by people it should map the values onto English phrases like "probably" or "almost certainly". This would be similar to the "linguistic hedges" of Schirra et al [61], and to the "fuzzy confidence measures" of Keller et al [42]. This step, however, has not been implemented in SOO-PIN.

Most of the enhancements described have been incorporated in SOO-PIN and run on various test data. The belief values were found to propagate through the system in a fairly straightforward manner and the output accorded with intuitive expectations. For instance, with the input shown in the XFIG diagram Figure 6.4, and with the network initiated with the messages:

```
msg(road,create(id(road,3,bel),[]))
msg(road,create(id(road,1,bel),[]))
msg(road,create(id(road,2,bel),[]))
msg(inXn,create(id(inXn,1,bel),[]))
msg(road,create(id(road,4,bel),[]))
msg(car,create(id(car,3,bel(0.5,1.0)),[]))
msg(car,create(id(car,4,bel(0.8,1.0)),[]))
```

ie, with the two cars given support of 0.5 and 0.8 respectively, then the system responded with:

```
id(carInXn,11194,bel(0.8,1.0))
id(carInXn,11192,bel(0.5,1.0))

id(straight,11193,bel(0.49,0.99)),reln(composedOf,id(car,3,bel(0.5,1.0)))

id(right,11195,bel(0.78,0.98)),reln(composedOf,id(car,4,bel(0.8,1.0)))

id(giveWayToOnc,11196,bel(0.39,0.98))

Give Way to oncoming:  id(car,4,bel(0.8,1.0)) turning right from east
        gives way to id(car,3,bel(0.5,1.0)) from west :  bel(0.39,0.98)
```

ie, the concept-instances id(straight,11193) and id(right,11195) have support of just less than the cars they are composed of. This is due to the car's proximity to a boundary in the intersection position-heading 3D product space (see Section 6.2) which explains why the plausibility is also slightly reduced. The concept-instance id(giveWayToOnc,11196) has support of 0.39 because this depends on the two activities (right and straight) it is composed of, and its support is the product of the supports of those activities (similarly for plausibility). The final interpretation has the same support as the id(giveWayToOnc) concept-instance as it depends on no other data.

The extra processing time to convey and calculate these uncertainty values is negligible.

7.5 Summary

In this chapter some approaches to uncertainty have been discussed, and how uncertainty has been incorporated into SOO-PIN. It is noted that the system adopted uncertainty in a fairly natural manner. In fact, it simplified the problem of dealing with negative information. The result of our implementation is a system that propagates uncertainty from the input objects and the spatial procedural subroutines in an efficient and straightforward manner, culminating in intuitively reasonable and useful output belief values.

Chapter 8

Velocity

8.1 Introduction

Earlier, in Chapter 6, it was found that SOO-PIN made mistakes in some situations that could be avoided if knowledge of vehicle velocities were available. In this chapter we deal with enhancements to the traffic scenario that allow the system to determine velocities. Having velocity gives the system the ability to discriminate the timing of events and the motion direction of vehicles. Thus it can determine whether cars are on a collision course, and it can determine what activity a car is performing on the road more accurately than from the orientation of the car alone. Knowing velocity allows questions like "why is car A stopped". Image sequences, from which velocity is found, gives the system more confidence in vehicle identification, and also, the tense of the English language output can be made to reflect more accurately what cars are doing. Note that in this chapter the examples incorporate the uncertainty calculus discussed earlier.

Velocity can be obtained from video by optical flow techniques, by finding correspondence between tokens in successive frames, or an amalgam of the two[1][15]. Since the original source of the data here is video, optical flow is a possible approach. However, Verri and Poggio in their paper "Against Quantitative Optical Flow" [66], argue that the optical flow and the motion field are in general different, and where they do correspond (when the image gradient is strong), feature based (ie, token) matching is more appropriate.

Mohnhaupt and Neumann[48], in a variation on optical flow, demonstrate a system in which traffic activity is found using top-down control of Gabor filter banks over a 4D phase space defined from image sequences. The Gabor filter banks output "energy" maps corresponding to the various vehicle activities, which, however, still need processing to generate vehicle "tokens".

Cipolla and Yamamoto [15] demonstrate a system that lies between optical flow and token matching which performs visual tracking in time and space using stereo video cameras. In this system epipolar lines in each image are extended into epipolar planes in the temporal domain using the video rate sequence of images. In this plane, moving objects form loci that are clearly distinguished from the fixed background. Thus object trajectories are found using edge detection and line completion in the spatio-temporal domain. The resulting trajectories are then matched between the left and right cameras using spatial and temporal continuity constraints, whereupon the depth can be inferred at each instance by trigonometry. Unfortunately this technique involves discarding a spatial dimension in order to focus on the epipolar line, which is not applicable to the 2D aerial view we have. It also requires dense video and fixed stereo cameras. Our system described below is based on sparse image sequences and fixed single camera. The 3D geometry is known directly from our assumption that the traffic is constrained to the road surface.

SOO-PIN uses token correspondence to track cars in the image sequence. This decision was taken because the input to the system consists of temporally sparse image sets, eliminating optical flow techniques, and was influenced by the fact that the system was already finding the tokens (ie, cars) in the images in the earlier stages of this project. It was also influenced by the existence of strong dynamical constraints imposed on moving vehicles by inertia, which can be easily exploited by the correspondence algorithm.

Below we describe how the correspondence problem is solved in our system, then how the network SOO-PIN is altered to accommodate velocity calculation, how the presence of velocity affects other parts of the network, and finally, how the output is improved with this extra information.

8.2   Finding Trajectories

Velocity is found by comparing three video frames separated by 400 milliseconds, as shown in Figure 8.1.

Although two frames are theoretically sufficient to calculate velocity, the system uses three frames (for example, see Figure 8.1) to facilitate the algorithm which is designed to reduce ambiguity in matching, and increase confidence.

The matching is based upon the centroids and orientations of the cars, how these are found in images is described in Section 6.5.

The middle frame is considered the master frame where each car found in the middle frame is compared with all cars in the other two frames. A pairing is successful if the apparent velocity required to move the first car to the second is reasonable for city traffic, and if this velocity is consistent with the orientations of the two cars as determined from their major axes (see the algorithm 8.2 below). The set of possible pairings between the middle and previous frames is compared with that from the middle and successor frames, and the best pair of pairs is chosen as the likely trajectory of the car.

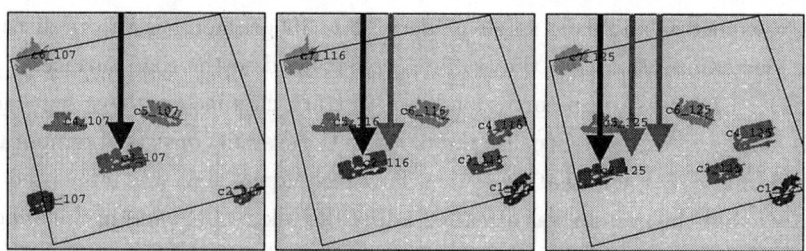

Figure 8.1: *Example of 3 successive frames with the movement of one car indicated by the arrows, faint arrows referring to previous car positions. The intersection boundary is shown by the skew rectangle, the car numbers are shown next to the cars.*

8.2.1 Matching between Frames

Here we describe how pairs of cars from successive frames are found. In the next section we describe how these pairs are matched to form a trajectory. The description begins with the pairing algorithm shown in Figure 8.2 which is demonstrated in Figure 8.3 for a moving car shown in two successive frames.

```
for each car p in P,
   for each car n in N,
      calculate velocity V: p → n
         and direction of velocity D degrees
      if |V| < min. detectable vel. /*car stopped*/
         if |orientation p - orientation n| is close to zero
            then accept match
      else if |V| < max. likely vel. /*car moving*/
         if |D - orientation p| < max rotation of car
            and |D - orientation n| < max rotation of car
               then accept match
   endfor
endfor
```

Figure 8.2: *Algorithm for finding matches between cars in successive frames. P refers to a frame, and N refers to its successor.*

The velocity vector (say, V) required to move a car between the two frames is first checked for its magnitude – that it did not exceed reasonable bounds for cars in intersections. Then that the orientations of the cars correspond to the direction of the velocity V, up to the limit shown by the arcs. The other case, where the car is stopped, is checked by determining whether the orientations correspond directly with each other. If a match is successful for, say, frames 1 and 2 of a triple, then the pair is added to the output list. This is passed back to the system for comparison with the match-list produced from frames 2 and 3.

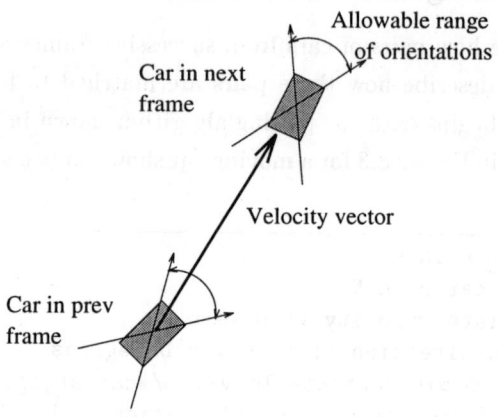

Figure 8.3: *A pair of cars in successive frames, showing how the velocity vector between the cars must be compatible with the orientations of each car. This compatibility is checked in Algorithm 8.2*

8.2.2   Finding Trajectories by Comparing Match Lists

The previous (frames 1 and 2) and next (frames 2 and 3) match lists found using the algorithm in Figure 8.2 are compared to find the most likely trajectories for each car in the middle frame. This is done by assigning a score to each possible trajectory based upon the difference between the speeds in the two matches, the difference between the rotation rates, and how the rotation derived from the two match velocities compares with the rotations of the matches themselves. Then the best score is selected as the likely trajectory of the car. This technique resolves a number of potential ambiguities discussed below.

In Figure 8.4 the match P2-M-N was rejected because the velocity P2-M is different to the velocities M-N and P1-M, and the system tries to minimize changes in velocity. In Figure 8.5 three cars are shown turning right. The system has a choice in dealing with car c2_20, it could correctly identify its trajectory as a right turner, or join it with c1_10 and c3_30, producing a car going straight. This option was rejected because of the skew orientations and the concordant rotation rates of the cars. If however the input is limited to the cars c1_10, c2_20 and c3_30 then the system would accept the straight trajectory because the algorithm is designed to find the best available, within bounds.

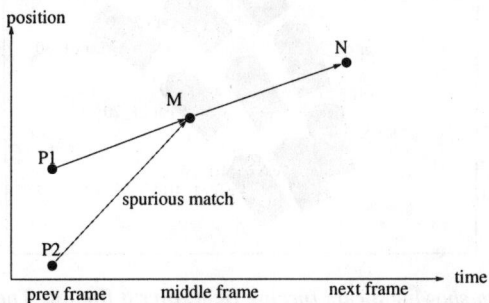

Figure 8.4: *Diagram showing a potential bad match (P2-M-N) and the correct match (P1-M-N), where the dots represent the positions of cars in 3 successive frames.*

8.2.3 Determining Velocity

Velocity determination is performed by the new concept-frame traj. This concept-frame is handed all car instances from the middle frame, and calls a C procedure (velocity.c) which returns lists of pairings from successor and previous frames. The lists are then compared in the Parlog++ language to determine the most likely trajectory of the car. This division of labour between C and Parlog++ is an instance of the system's basic design philosophy, in which spatial predicates are handled by C, and more symbolic processes are handled by Parlog++.

The chosen trajectory is then stored as an instance of the traj concept-frame, the corresponding velocity is sent to the originating car concept-frame, and check messages are sent higher in the network to carInXn, carTInXn and carRoad which initiate further processes, Figure 8.6 shows how the new concept-frame relates to the rest of the network.

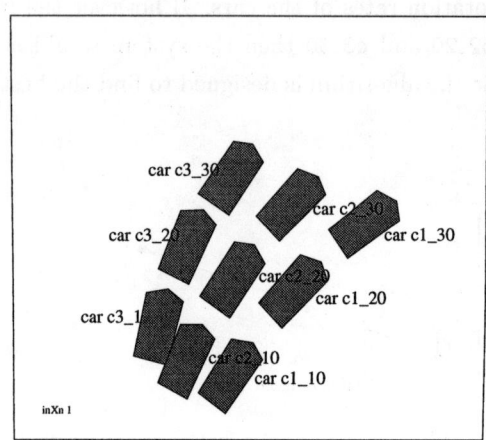

Figure 8.5: *Diagram showing 3 cars turning right, over 3 frames. The system successfully finds the 3 trajectories.*

8.3   Uses of Velocity

As can be seen from Figure 8.6, a message is also sent from `traj` to `collision`. This concept-frame determines whether the car in the message is on a collision course with any other car. This is performed by calling a C routine that uses the velocities and positions of each pair of cars to calculate if the trajectories (nearly) intersect within a set period of time (see Figure 8.7). If so, a collision warning message is output.

The concept-frames `carInXn`, `carTInXn` and `carRoad` get velocity data from the `car` concept-frame, and use this velocity rather than orientation (as previously) to calculate the turn activity of the car in the intersection (or road). This increases the reliability of the resulting activity type. It also means that if a car is on the wrong side of the road, the system detects this rather than assuming (as before) that the car is traveling in the opposite direction. This generates a new kind of high level output from the system: declarations of illegality status like "car on wrong side of road".

If the concept-frames `carInXn`, `carTInXn` and `carRoad` do not find velocity

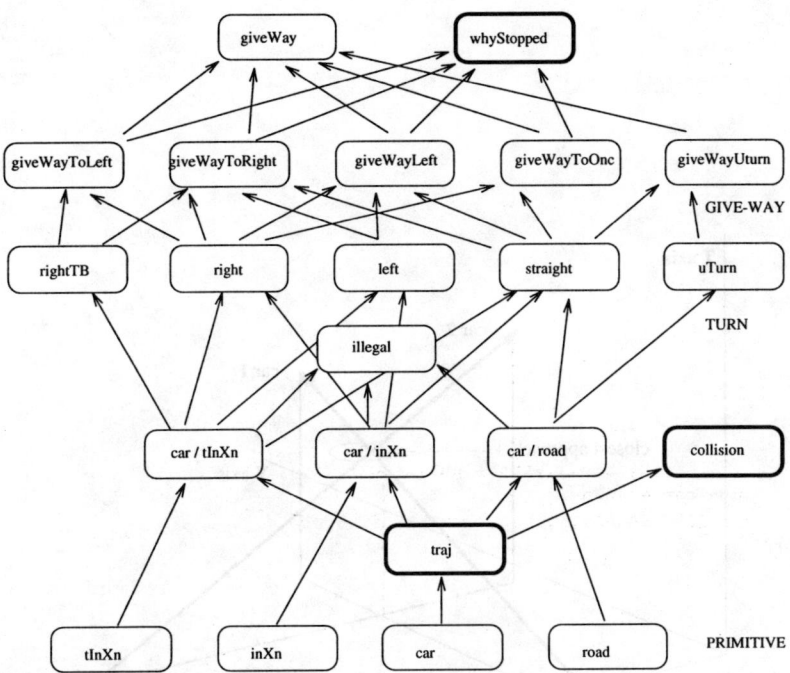

Figure 8.6: *The expanded network with the concept-frames dealing with velocity shown in bold. For comparison see Figure 6.2.*

for a given car (ie, the car is not matched with cars in other frames) then they have two possible responses:

- ignore the car (since it is not confirmed in other frames, it may not exist);

- use its orientation alone and continue, albeit with a lower "belief value" for subsequent deductions. This option also generates less precise interpretations as described below.

SOO-PIN uses the second option, as it produces more opportunities for further interpretation.

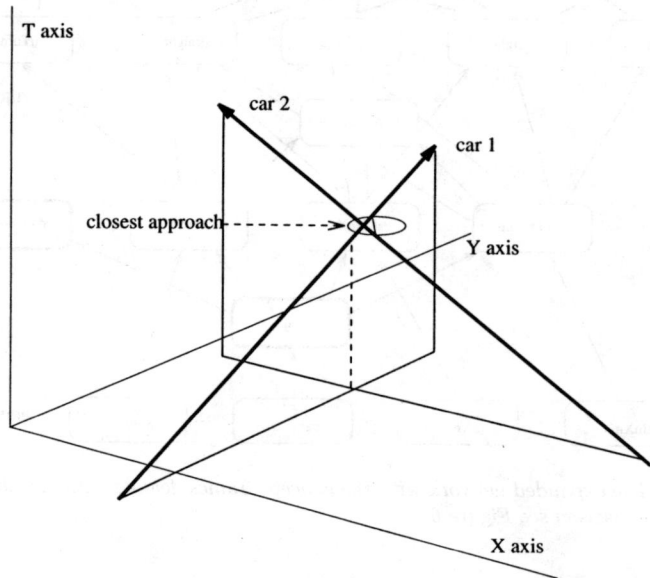

Figure 8.7: *The geometry involved in calculating collisions. The trajectory of each car is given by dark arrows in a 3D space of X and Y coordinates on the ground, and the time T axis vertically. The cars are assumed to be in danger of collision if the closest approach (parallel to the X-Y plane) of the two trajectories is less than a threshold (whose value is set to the size of the average vehicle in the image) within a short period (set to be approximately the reaction time of the average driver).*

If a give-way concept-frame does not create an instance for a given car, it sends a check message to whyStopped. This concept-frame first checks the velocity of the car, and if it is below a threshold, regards the vehicle as stopped and sets about finding a reason for it. This is done by sending a message to all other give-way concept-frames to see if the car is involved in some other give-way relationship. If not, a message is sent to the output suggesting why the car may be stopped: for instance, an undetected car or a pedestrian. This concept-frame could be extended to search more intelligently for a reason why the car is stopped, for instance, if a car in front is also stopped and blocking the first car's way. Alternatively, it could start a resegmentation in the low-level vision system in the target area in an attempt to find an obstacle (but that goes beyond the scope of the current project).

The final place where velocities are useful is in generating the English language interpretations. For instance, before velocity was available the system produced:

```
*Give Way to oncoming:  id(car,c5_116,bel(1,1))
        turning right from west gives way to
        id(car,c2_116,bel(1,1)) from east :  bel(0.98,0.98)
```

That is, the system could detect that there was a give-way *situation*, but it had no way of determining the tense, ie, whether id(car,c5_116,bel(1,1)) had given way, was giving way, or had not yet given way. With velocity this distinction can be clarified, producing output like:

```
*Give Way to oncoming:  id(car,c5_116,bel(1,1))
        turning right from west has given way to
        id(car,c3_116,bel(1,1)) from east :  bel(0.98,0.98)
```

and

```
*Give Way to left-turner:  id(car,c6_116,bel(1,1))
        turning right from west should have (but hasn't) given way to
        id(car,c4_116,bel(1,1)) from east :   bel(0.96,0.96)
```

Given acceleration, the system can also generate phrases like "is giving way to"; moreover, acceleration is potentially available from the traj concept-frame since it has two velocities from comparing previous and successor frames. However, given the unreliability of the starting positional data, the reliability of its second derivative (acceleration) is far worse, so it is not used.

8.4   Velocity Examples

The first example is given in Figure 8.8, in which 4 cars are shown in 3 frames, the velocity determination system had to find which of c1_10 and c2_10 was the best fit for the car trajectory. The output from the concept-frame `traj` (where velocity determination takes place) was as follows:

```
*adding inst 11211 composed of id(car,c1_20,bel) and
       [id(car,c1_30,bel),id(car,c1_10,bel)] with
       vel(0.05,-9.55) and rotation(0.0)

*inquiry to id(traj,11211,bel(0.94,1)) re desc(vel(any,any),bel)
```

As can be seen, the system matched the cars id(car,c1_20,bel), id(car,-c1_30,bel), id(car,c1_10,bel) to construct the trajectory, avoiding the bad match id(car,c2_10,bel), which is based on the supposition that the speed of the car changes slowly. The velocity so determined was vel(0.05, -9.55), meaning 9.55 pixels per frame vertically, and 0.05 pixels per frame to the right. The rotation rate was zero. It can be seen that subsequently a message was received by `traj` from another concept-frame inquiring about this velocity.

The next example, shown in Figure 8.9, demonstrates the system choosing a trajectory that is consistent in curvature as opposed to an S-shaped trajectory. This situation arises if the low-level system returns regions from a noisy image. The system has to construct the most sensible trajectory, guided by some physical principles, for instance:

- cars do not exceed a threshold speed

- cars do not exceed a threshold acceleration

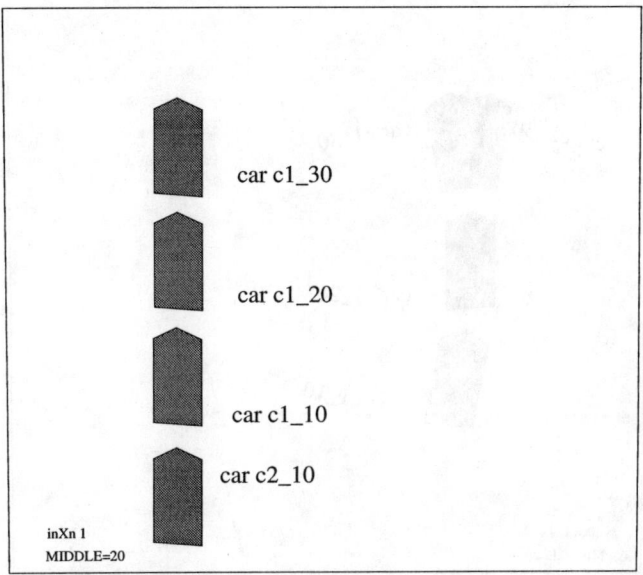

Figure 8.8: *Example of velocity from 4 cars in 3 frames, in which the correspondence could be confused between cars* c1_10 *and* c2_10. *Here all 3 frames are shown in same diagram, the frame numbers are determined by the system from the last number in the car identity string.*

- cars do not exceed a threshold rotation

- cars do not exceed a threshold change in rotation

In this case, the principle is the last, ie, if the car is rotating, it will continue to rotate for some time. Thus the output from the traj concept-frame was as follows:

```
*adding inst 11244 composed of id(car,c1_20,bel) and [id(car,c1_30,bel),
     id(car,c1_10,bel)] with vel(-0.0125,-9.86) and rotation(0.96)

*inquiry to id(traj,11244,bel(0.93,1)) re desc(vel(any,any),bel)
```

It can be seen that the system selected the constant curvature trajectory, with a velocity of about 10 pixels per frame to the north and a rotation of about

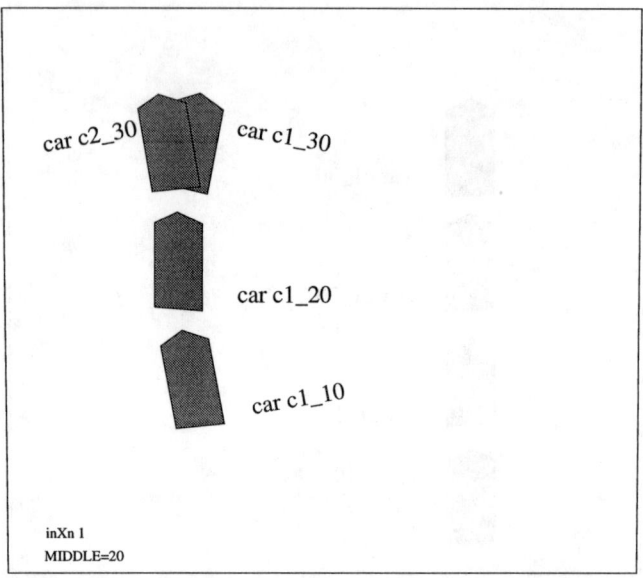

Figure 8.9: *Example of velocity from 4 cars in 3 frames, in which the correspondence could be confused between cars* c2_30 *and* c1_30. *Here all 3 frames are shown in same diagram, the frame numbers are determined by the system from the last number in the car identity string.*

1 degree per frame clockwise.

The last velocity example (Figure 8.10) demonstrates the collision detection system, employing the concept-frame `collide`. One car is traveling "illegally" from the east, the other traveling north. The system interpretation for this configuration is:

```
*Collision imminent:  id(car,c2_20,bel) is on collision course
     with id(car,c1_20,bel) :  bel(0.83,0.92)

*Illegal:  car id(car,c2_20,bel(1,1)) from east is on the wrong
     side of intersection id(inXn,1,bel) :  bel(0.93,0.99)
```

The collision warning and the illegality status was correctly identified, the belief values are also shown for these assertions. The output from the `traj`

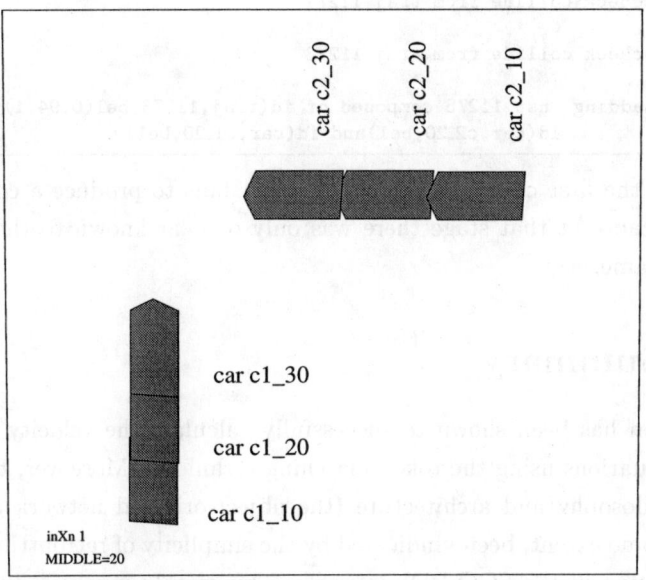

Figure 8.10: *Example of detection of a collision between two cars, one traveling "illegally" from the east, the other traveling north. Here all 3 frames are shown in same diagram, the frame numbers are determined by the system from the last number in the car identity string.*

concept-frame for this configuration is:

```
*adding inst 11271 composed of id(car,c1_20,bel) and
        [id(car,c1_30,bel),id(car,c1_10,bel)] with
        vel(0.05,-5.3) and rotation(0.0)

*inquiry to id(traj,11271,bel(0.94,1)) re desc(vel(any,any),bel)

*adding inst 11273 composed of id(car,c2_20,bel) and
        [id(car,c2_30,bel),id(car,c2_10,bel)] with
        vel(-7.55,-0.05) and rotation(0.0)

*inquiry to id(traj,11273,bel(0.94,1)) re desc(vel(any,any),bel)
```

and finally, the output from the collide concept-frame is:

```
*check collide from traj 11271

*check collide from traj 11273

*adding inst 11275 composed of id(traj,11273,bel(0.94,1))
          id(car,c2_20,bel)and id(car,c1_20,bel)
```

Note how the first check message from traj fails to produce a collision instance because at that stage there was only one car known to the collide concept-frame.

8.5   Summary

The system has been shown to successfully calculate the velocity of cars in certain situations using the token matching technique. Moreover, the overall design philosophy and architecture (the object-oriented network approach) have, to some extent, been vindicated by the simplicity of retrofitting velocity determination in the SOO-PIN network. This simply involved building the new concept-frame traj, together with its C subroutines, and patching it into the network by redirecting some messages (in this case, redirecting car check messages from carInXn carRoad and carTinXn to the new traj and directing traj output to those concept-frames. Other concept-frames were then able to use velocity by sending inquiry messages to car or traj.

This chapter has shown how knowledge of vehicle velocity allows better interpretations of the scene, for instance, more precise statements about the interactions of the cars. Knowledge of velocity has also allowed new interpretations, like whether cars are on a collision course, or whether a car is on the wrong side of the road. Finally, velocity information, and the correspondence between frames upon which it is based, improves confidence in the vehicle identifications and thus in the output interpretations.

Chapter 9

Runtime Results

9.1 Introduction

In this chapter runtime results are presented and discussed. The input consists of XFIG diagrams and selected frames of video from real traffic intersections. For real images, the intermediate segmented and labeled images are also shown. The output, being the automatic interpretation, is shown in the figure captions.

9.2 Trial Runs

The first example in Figure 9.1 is of a complex situation in a T-intersection created using the XFIG drawing package. This example is intended to demonstrate the capabilities of SOO-PIN using situations that are rarely found in video footage of real traffic. The first interpretation shows that the system successfully found that car c4_89 was illegally placed, as it was turning right into the blank side of the T-intersection. The next interpretation demonstrates the give-way concept-frame making use of velocity information. Here, it found that c1_89 was still moving when it was facing a red light. Similarly, for c3_89, although this case is interesting because c6_89 was not a member of a triple, and hence has no velocity information. This resulted in its existence being more uncertain than the other cars, and thus the interpretation has a

103

lower support, namely 0.36. This is also true of the interpretations involving cars c1_89 and c6_89. The last interesting interpretation concerns c2_89, which was traveling slowly enough for the system to classify it as stopped. The whyStopped concept-frame was activated when the system could not find a reason for the car being stopped, in this case c2_89 would have been waiting for c1_89 to pass by, but the system was not complex enough to handle that case. Note that this example is similar to that of Figure 6.6, but in this case including multiple frames and thus velocity. Comparing the interpretations from these two cases, it is found that the latter case has more precise give-way verbs, and the additional interpretation regarding why car c2_89 was stopped.

The next example, Figure 9.2, is of an intersection also created using the XFIG drawing package. The first interpretation demonstrates how having velocity allowed the system to determine whether the car was traveling on the wrong side of the road. There were also a number of cars supported by only two frames, which demonstrate how the system robustly uses the information available to it in order to calculate velocity, albeit with lower certainty (see cars c5_89 and c7_89). This example also demonstrates another whyStopped interpretation.

Figure 9.3 shows a real image and it's segmented and labeled counterpart. It can be seen that the system had identified the activity of each car and interpreted every interaction correctly. Note how car c5_89 had not interacted with c7_89, as this car had already passed by. This is an example of the spatial "targeting" of cars being taken into account in the give-way concept-frames, meaning that a car only gives way to another if the cars have the correct spatial orientation with respect to each other. Also note that the verb used in the interpretation is "has given way to". This is because the car that gives-way was stopped. It can be seen from their long velocity arrows that the four cars c1_89, c3_89, c6_89 and c7_89 were moving rapidly through the intersection , car c5_89 was moving slowly, from its short but correctly oriented arrow, and the rest of the cars have zero-length randomly directed arrows and were thus stopped.

Figure 9.4 shows the same scene as Figure 9.3 but one second later, the interactions are qualitatively similar. Note how the dark car visible in the top right corner of the intersection, as noted in Section 6.5, had escaped the attention of the system as its colour was similar to the background asphalt. The same car was identified in Figure 9.3 because, at that time, it was visible against the lighter coloured concrete tramway. Here a more sophisticated low-level object recognition system would have been useful. Compare the interpretation of this scene with that in Section 6.6 (Figure 6.8) which is based on the same image but in that case without velocity. Having velocity, the van c4_116 was correctly identified as a straight-througher, whereas earlier it was identified as a left-turner due to its skewed appearance. This example demonstrates the importance of determining car velocity.

Figure 9.5 shows an example of the low-level processing coming to grief, the two cars from the north have both been segmented into two parts, resulting in a number of redundant interpretations, one for each part, and culminating in the collision detector being activated for some of these parts.

Figure 9.6 shows yet another segmentation problem, here the left-turner (c2_130 and c3_130) had been split into two by the lamp-post. One half was still correctly interpreted as a left-turner, the other half appeared, to the system, to be turning right. Based on this false premise, the system had drawn a reasonable conclusion. This image also draws attention to a possible improvement to the traffic implementation of SOO-PIN, the right-turning car c4_130 was waiting for the car in front of it, a concept-frame that looked for such situations and reported them would have been a nice extension.

Figure 9.7 shows the same scene as Figure 9.6 but nearly one second later. The interpretations this time were correct.

Figure 9.8 is again the same scene as above but nearly two seconds later when the lights have changed. This is a very simple scene that had been correctly interpreted.

Figure 9.9 shows a T-intersection taken from an oblique angle. Here car c1_120 was waiting for car c2_120, traveling on the through road, to clear before turning right.

Figure 9.10 shows the same intersection as above some time later. This interpretation is unusual in that all cars were seen to be moving, and hence the system generated the "should have (but hasn't) given way" give-way verb in the interpretation of the scene.

Figure 9.11 shows the only "illegal act" caught on the video. Car c5_316 was doing a U-turn in the T-intersection. The system deduced that the car was doing an illegal right-turn, which was the only interpretation available to it under the circumstances. This figure also shows a problem with oblique viewing angles, the "car" c1_316 is actually two cars which were merged due to the roof of one car adjoining the shadow of the second. Again, this draws attention to the need for a more sophisticated low-level processing system if SOO-PIN were to be implemented as an industrial product.

9.3   Summary of Results

In the examples above, the system was run on 9 scenes, involving 27 images, from 3 different intersection. This resulted in 19 correct interpretations and 4 erroneous interpretations (ie, 82% correct), as judged by visual inspection. However, if the interpretations are weighted by the support given in their belief pair, the accuracy increases to 89%. This indicates that wrong interpretations have lower than average support, in other words, that our belief calculus is effective.

All of these wrong interpretations involved cars that were split in two due to a section of the car merging with the background or being obscured by foreground (see Section 6.5). These problems could be avoided using a better low-level object recognition system, for instance, the model-based system described by Boddington ([10]), or by utilizing domain knowledge through incorporating the low-level system into the network-of-frames.

On an SGI Personal Iris with one 33MHz IP12 processor these 9 scenes took 1763 seconds of CPU time for low level processing, and 76 seconds of CPU on a Sun SparcStation II for the high level (Parlog++) processing, ie, just over 3 minutes per scene. This was achieved with no attempt at fine tuning the system for speed. With a more sophisticated low level processor we could eliminate some of the errors, and achieve higher processing speed.

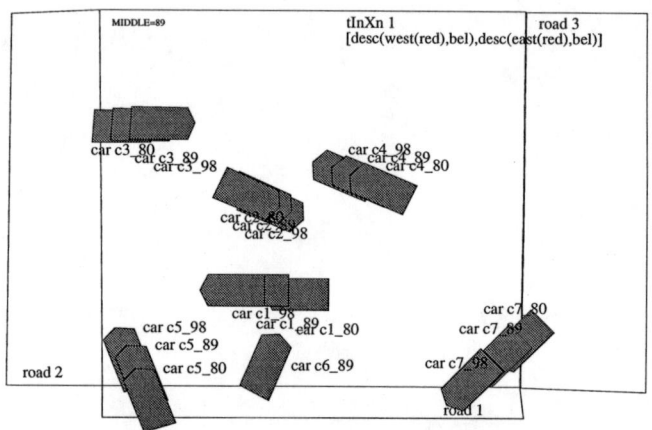

Figure 9.1: *Cars in T-intersection, XFIG diagram of unrealistic situation. The cars from all three frames are shown in the one image, the system determines the frame number from the car identity string. North is up, the traffic lights are given with the intersection name. The interpretation follows:*

∗Illegal: car id(car,c4_89,bel(1,1)) from east is on the wrong side of T-intersection id(tInXn,1,bel) : bel(0.93,0.99)

∗Give Way to oncoming overridden by Traffic Sign red : id(car,c1_89,bel(1,1)) going straight from east should have (but hasn't) given way to id(car,c2_89,bel(1,1)) turning right from west : bel(0.57,0.92)

∗Give Way to Left (T-inXn) overridden by Traffic Sign red : id(car,c3_89,bel(1,1)) from west should have (but hasn't) given way to id(car,c6_89,bel(1,1)) from south : bel(0.36,0.97)

∗Give Way overridden by Traffic Sign red : id(car,c1_89,bel(1,1)) from east should have (but hasn't) given way to id(car,c5_89,bel(1,1)) from south : bel(0.56,0.93)

∗Give Way to Right overridden by Traffic Sign red : id(car,c1_89,bel(1,1)) from east should have (but hasn't) given way to id(car,c6_89,bel(1,1)) from south : bel(0.24,0.93)

∗Give Way to left-turner: id(car,c2_89,bel(1,1)) turning right from west has given way to id(car,c7_89,bel(1,1)) from east : bel(0.60,0.98)

∗id(car,c2_89,bel(1,1)) turning right is stopped : It may be waiting upon a car coming from the east , or upon pedestrians crossing on the south side: bel(0.90,0.98)

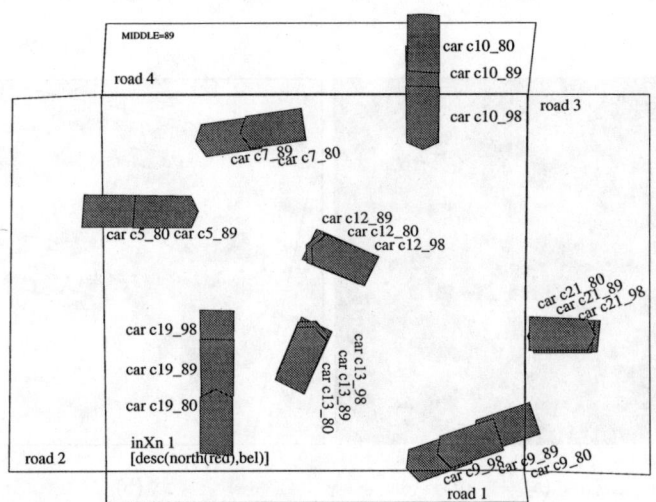

Figure 9.2: *Cars in intersection, XFIG diagram of unrealistic situation. The cars from all three frames are shown in the one image, the system determines the frame number from the car identity string. North is up, the traffic lights are given with the intersection name. The interpretation follows:*

*Illegal: car id(car,c7_89,bel(1,1)) from north is on the wrong side of intersection id(inXn,1,bel) : bel(0.63,0.97)

*Give Way to oncoming: id(car,c13_89,bel(1,1)) turning right from south has given way to id(car,c10_89,bel(1,1)) from north : bel(0.61,0.98)

*Stop for traffic sign: id(car,c10_89,bel(1,1)) on id(road,4,bel) should have (but hasn't) stopped for traffic sign red : bel(0.64,0.99)

*Give Way to Right: id(car,c13_89,bel(1,1)) from south has given way to id(car,c21_89,bel(1,1)) from east : bel(0.92,0.97)

*Give Way to Right: id(car,c13_89,bel(1,1)) from south has given way to id(car,c12_89,bel(1,1)) from east : bel(0.93,0.98)

*id(car,c21_89,bel(1,1)) going straight is stopped : It may be waiting upon a car in the next intersection, or upon a traffic sign: bel(0.93,0.98)

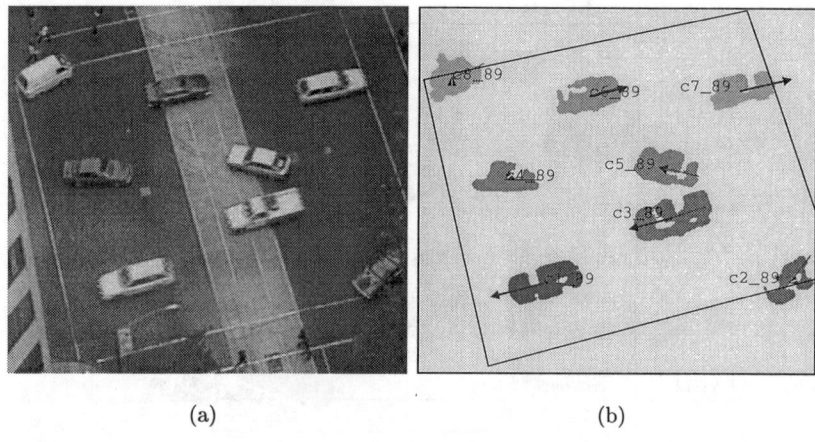

(a)   (b)

Figure 9.3: *Collins & Exhibition Sts., frame 89. (a) Cars in intersection, middle image of the triple. (b) Cars found by system with labels attached, and velocity given by arrow lengths. The intersection boundary is input manually, north is up. The interpretation follows:*

*Give Way to left-turner: id(car,c4_89,bel(0.86,1.0)) turning right from west has given way to id(car,c2_89,bel(0.69,1.0)) from east : bel(0.57,0.98)

*Give Way to oncoming: id(car,c4_89,bel(0.86,1.0)) turning right from west has given way to id(car,c3_89,bel(0.97,1.0)) from east : bel(0.78,0.98)

*Give Way to oncoming: id(car,c4_89,bel(0.86,1.0)) turning right from west has given way to id(car,c1_89,bel(1.0,1.0)) from east : bel(0.82,0.98)

*Give Way to oncoming: id(car,c5_89,bel(1.0,1.0)) turning right from east has given way to id(car,c6_89,bel(1.0,1.0)) from west : bel(0.74,0.97)

*Give Way to left-turner: id(car,c5_89,bel(1.0,1.0)) turning right from east has given way to id(car,c8_89,bel(0.78,1.0)) from west : bel(0.51,0.97)

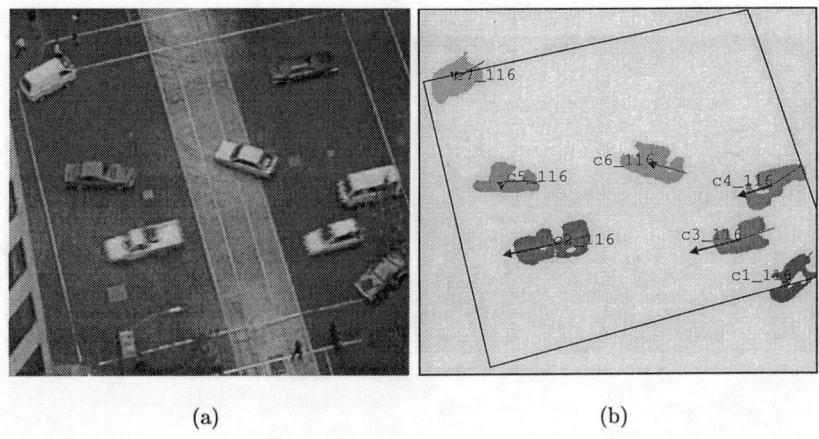

(a)   (b)

Figure 9.4: *Collins & Exhibition Sts., frame 116. (a) Cars in intersection, original image.*
(b) Cars found by system, with orientations given by lines through cars, and their labels.
Intersection boundary input manually, north is up. The interpretation follows:
  *Give Way to oncoming: id(car,c5_116,bel(1,1)) turning right from west has given
way to id(car,c4_116,bel(1,1)) from east : bel(0.60,0.98)
  *Give Way to oncoming: id(car,c5_116,bel(1,1)) turning right from west has given
way to id(car,c3_116,bel(1,1)) from east : bel(0.62,0.98)
  *Give Way to oncoming: id(car,c5_116,bel(1,1)) turning right from west has given
way to id(car,c2_116,bel(1,1)) from east : bel(0.90,0.98)
  *Give Way to left-turner: id(car,c5_116,bel(1,1)) turning right from west has given
way to id(car,c1_116,bel(1,1)) from east : bel(0.93,0.98)
  *Give Way to left-turner: id(car,c6_116,bel(1,1)) turning right from east has given
way to id(car,c7_116,bel(1,1)) from west : bel(0.93,0.99)

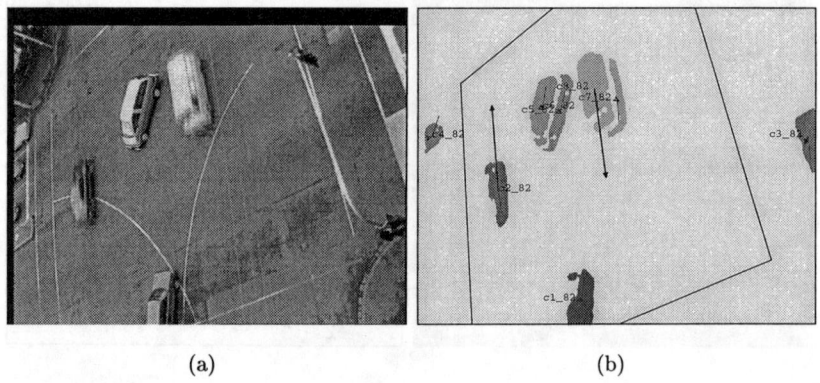

(a)   (b)

Figure 9.5: *Lygon & Queensberry Sts., frame 82. (a) Cars in intersection, original image.*
(b) Cars found by system, with orientations given by lines through cars, and their labels.
Intersection boundary input manually, north is up. The interpretation follows:
    *Give Way to left-turner: id(car,c4_89,bel(1,1)) turning right from west has given way
to id(car,c2_89,bel(1,1)) from east : bel(0.93,0.98)
    *Give Way to oncoming: id(car,c4_89,bel(1,1)) turning right from west has given way
to id(car,c3_89,bel(1,1)) from east : bel(0.90,0.98)
    *Give Way to oncoming: id(car,c4_89,bel(1,1)) turning right from west has given way
to id(car,c1_89,bel(1,1)) from east : bel(0.90,0.98)
    *Give Way to oncoming: id(car,c5_89,bel(1,1)) turning right from east has given way
to id(car,c6_89,bel(1,1)) from west : bel(0.89,0.97)
    *Give Way to left-turner: id(car,c5_89,bel(1,1)) turning right from east has given way
to id(car,c8_89,bel(1,1)) from west : bel(0.62,0.97)

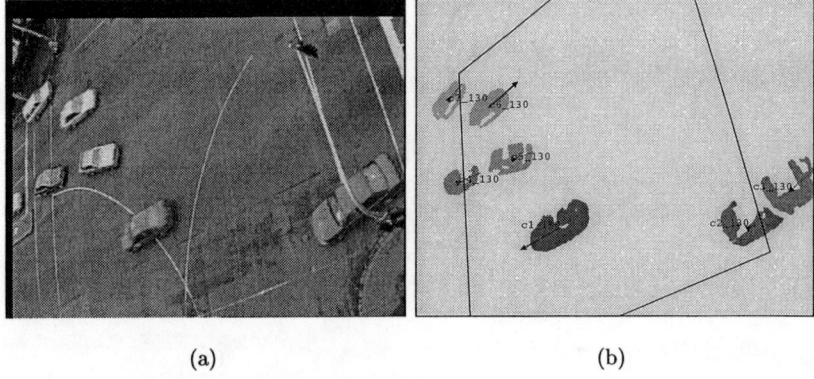

(a)   (b)

Figure 9.6: *Lygon & Queensberry Sts., frame 130. (a) Cars in intersection, original image.*
(b) Cars found by system, with orientations given by lines through cars, and their labels.
Intersection boundary input manually, north is up. The interpretation follows:
  *Give Way to oncoming: id(car,c5_130,bel(1,1)) turning right from west has given
  way to id(car,c1_130,bel(1,1)) from east : bel(0.61,0.98)
  *Give Way to oncoming: id(car,c2_130,bel(1,1)) turning right from east has given way
  to id(car,c6_130,bel(1,1)) from west : bel(0.39,0.92)

(a)   (b)

Figure 9.7: *Lygon & Queensberry Sts., frame 150. (a) Cars in intersection, original image.
(b) Cars found by system, with orientations given by lines through cars, and their labels.
Intersection boundary input manually, north is up. The interpretation follows:*
   *Give Way to left-turner: id(car,c4_150,bel(1,1)) turning right from west has given
   way to id(car,c1_150,bel(1,1)) from east : bel(0.62,0.98)
   *Give Way to left-turner: id(car,c5_150,bel(1,1)) turning right from west has given
   way to id(car,c1_150,bel(1,1)) from east : bel(0.62,0.98)

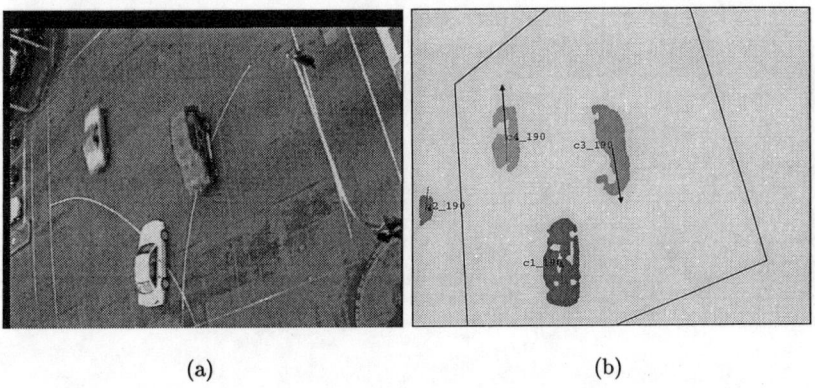

(a)   (b)

Figure 9.8: *Lygon & Queensberry Sts., frame 190. (a) Cars in intersection, original image. (b) Cars found by system, with orientations given by lines through cars, and their labels. Intersection boundary input manually, north is up. The interpretation follows:*
  *Give Way to oncoming: id(car,c1_190,bel(1,1)) turning right from south has given way to id(car,c3_190,bel(1,1)) from north : bel(0.61,0.98)

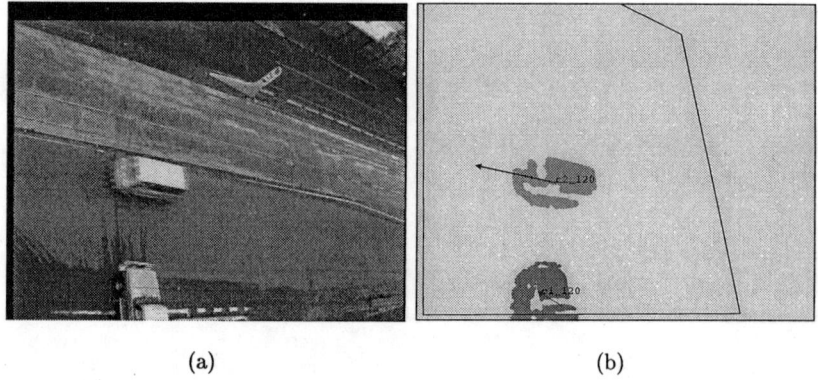

(a)   (b)

Figure 9.9: *Swanston & Faraday Sts., frame 120. (a) Cars in intersection, original image.*
(b) Cars found by system, with orientations given by lines through cars, and their labels.
Intersection boundary input manually, north is up. The interpretation follows:
   *Give Way: id(car,c1_120,bel(1,1)) from south has given way to id(car,c2_120,bel(1,1))
   from east : bel(0.59,0.98)

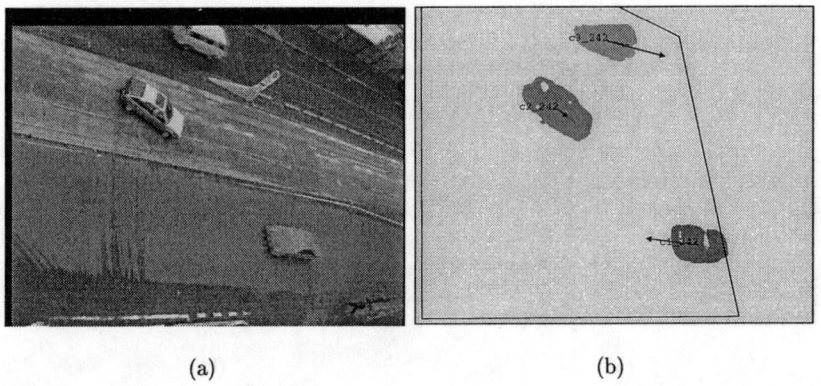

(a)   (b)

Figure 9.10: *Swanston & Faraday Sts., frame 242. (a) Cars in intersection, original image. (b) Cars found by system, with orientations given by lines through cars, and their labels. Intersection boundary input manually, north is up. The interpretation follows:*
  *Give Way to oncoming: id(car,c2_242,bel(1,1)) turning right from west should have (but hasn't) given way to id(car,c1_242,bel(1,1)) from east : bel(0.40,0.97)

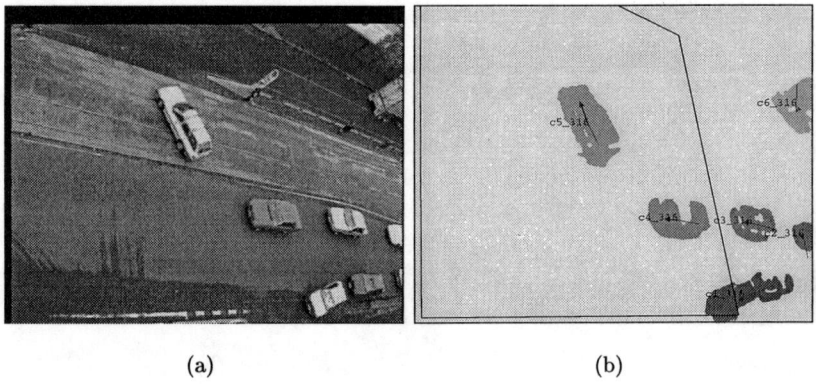

(a)   (b)

Figure 9.11: *Swanston & Faraday Sts., frame 316. (a) Cars in intersection, original image.*
(b) Cars found by system, with orientations given by lines through cars, and their labels.
Intersection boundary input manually, north is up. The interpretation follows:
  Illegal: car id(car,c5_316,bel(1,1)) from east is doing a right turn on the wrong side
  of id(tInXn,1,bel) : bel(0.72,0.78)

Chapter 10

Conclusion

The philosopher Dan Dennett has described in [25] his concept of how consciousness is a product of a network of interacting agents, none of which is the "center" or "seat" of consciousness. In this book we have demonstrated a system for interpreting images which, likewise, has no central controller. Rather it is a network of interacting (peer) agents, which, nevertheless, produces stories about images. This is a far cry, of course, from consciousness, but it does demonstrate how interesting high-level concepts can be generated from low-level data through networks of agents. In this book, the agents (concept-frames, or in Hewitt's language, actors) behave as Hewitt prescribed [38]. The relationship with Minsky's frames [47], is also close, but his slots (or terminals) are not explicitly implemented in SOO-PIN, their role is taken by property lists attached to each concept-instance. Together with the high-level logic programming language in which SOO-PIN is implemented, property lists provide a flexibility well suited for building a wide variety of concept-frames necessary in a rich domain. At an engineering level, the SOO-PIN system has demonstrated that an Object-Oriented Concurrent Logic Programming approach is a viable means of producing useful image interpretation systems. Another capability shown to be feasible by SOO-PIN is the use of a form of Dempster-Shafer uncertainty calculus, which was shown to convey useful uncertainty values through the network and produce intuitively reasonable uncertainty values at the final interpretation stage.

At the level of the traffic scenario, the system generated reasonable high level interpretations which would be suitable for input into a number of systems, for instance, quite sophisticated traffic intersection statistics could be accumulated, which would be of use to highway engineers. It would not take too much extension to connect such a traffic interpretation system to actual traffic light control and adjust the lights as a function of the types of driving behaviors detected. In fact, in conjunction with a street-level camera, the system could detect and report more complex illegal movements of cars as part of a traffic law infringement system.

Clearly, the present traffic implementation has many limitations. For instance, the background subtraction technique for low-level processing is crude, and results in problems with shadows and obscuration by overhanging structures, and limits the obliqueness of the view to the near vertical (see Figures 9.5 9.6 9.11). The system would be more robust with an intelligent object recognition system, such as one based on evidence-based techniques [13][8]. Output from such low-level processors should include uncertainty values which could be incorporated with the uncertainty measures currently propagated through SOO-PIN.

Improvements to the traffic implementation could also be made in the range of concept-frames. For instance, the system was not developed to detect whether a car was *intending* to turn left or right while still in the road next to an intersection. Also, the whyStopped concept-frame could be extended to reason that cars stop because the car in front is blocking the way. The system could also be extended to deal with other legally significant entities like pedestrians, trams and emergency vehicles.

The uncertainty measure propagated through the system could be improved by making the system more reactive to it, for instance, eliminating hypotheses that have a belief below a certain threshold. The belUpd message is a useful means of dealing with negative information (ie, the deletion of concept-instances) and of updating belief values around the network. This could have been implemented for the traffic scenario, except that, in some ways, the traffic scenario was quite simple, and there was no situations where hypotheses need to be deleted. Another improvement would be to map the uncertainty values onto English phrases (hedges) for output, and it would be an interesting problem to determine the most reasonable English phrases for the various combinations of support and plausibility output by the system.

Future work for the SOO-PIN system also includes extending the network of frames into low-level processing, creating concept-frames concerned with segments, image attributes, labeled regions and objects. With such a system, it would be possible to re-segment portions of the image where high-level concept-frames expect to find objects (ie, cars) while varying the segmentation parameters. This would be similar to the techniques used in Schema [27], Sigma [45] and that of Bell and Pau [7].

In the traffic scenario, SOO-PIN was working on sparse images derived from video image sequences. An obvious extension would be to deal with the sequence in its entirety. This would require a fundamental rethink of the architecture of the system, involving concepts of short-term and long-term memory, time-varying versus static objects, and a different form of output (whereas currently the system outputs interpretations when it has finished running, there would need to some way of determining when was the right time to generate output if it was running over a long sequence).

Finally, for the programmer, a useful improvement would be a graphical user interface (GUI) for building the system, similar to the system employed by Garvey [31] for his belief network. The current system was built using the high-level logic programming language Parlog++, which greatly facilitated the process, but it would be good to avoid the need for future programmers to learn this language. If there was a suitable GUI in which the logical nature and relationships of each concept-frame could be defined, then the Parlog++ code could be compiled from it. More speculatively, and in keeping with Cognitive Science, it would be a significant development if Machine-Learning could be involved in the construction (or evolution) of the network of concept-frames.

Appendix A

Parlog++ Procedures

In this appendix we list selected code in the Parlog++ language used in SOO-PIN.

A.1 Switchboard Source Code

The first source code example is the switchboard which controls message flow around the network, and spawns new concept-frames as needed.

```
/* PARLOG++ 'switchboard' that accepts messages on the input stream
   and routes them to the appropriate concept-frame.  If not found,
   it spawns the addressed concept-frame.  */

switch.
        InStr istream, WriStr ostream

invisible ProcList state <= [], ObjType state <= switch,
        MsgCount state <= 0

clauses

        InStr::msg(NameTo,Msg) =>
                find_name(ProcList,NameTo,ProcInput,PLshort):
                %debug, inq next line
                ProcInput = [Msg|ProcInput1]&
                WriStr::msg(NameTo,Msg)&          /*debug*/
                %writeMy(OutFile,msg(NameTo,Msg))&
                Count is MsgCount + 1,
```

```
            writeCount(Count)&
            MsgCount becomes Count&
            ProcList becomes [proc(NameTo,ProcInput1)|PLshort] ;

    InStr::msg(NameTo,Msg) =>
            checkProc(NameTo):
            Proc =.. [NameTo,[Msg|NewProcInput],NewProcOutput],
            call(Proc),           /*parallel, spin off new process*/
            WriStr::msg(NameTo,Msg),            /*debug*/
            %writeMy(OutFile,msg(NameTo,Msg))&

            merge(NewProcOutput,InStr,NewInStr),
            ProcList becomes [proc(NameTo,NewProcInput)|ProcList],
            Count is MsgCount + 1,
            writeCount(Count)&
            MsgCount becomes Count&
            InStr becomes NewInStr ;

    InStr::msg(NameTo,Msg) =>
            Count is MsgCount + 1,
            writeCount(Count)&
            MsgCount becomes Count&
            WriStr::[unknown_proc,NameTo,Msg]&& ;

    InStr::err(ErrMsg) =>
            Count is MsgCount + 1,
            writeCount(Count)&
            MsgCount becomes Count&
            WriStr::[error_rcvd,ErrMsg]&& ;

    InStr::info(InfoMsg) =>
            Count is MsgCount + 1,
            writeCount(Count)&
            MsgCount becomes Count&
            WriStr::[info_rcvd,InfoMsg]&& ;

    InStr::draw =>
            drawNet(ProcList,NewProcList)&
            WriStr::draw&
            Count is MsgCount + 1,
            writeCount(Count)&
            MsgCount becomes Count&
            ProcList becomes NewProcList&& ;

    InStr::quit =>
            WriStr::stopping_procs&
            WriStr::msgCount(MsgCount)&
```

```
                   stop_proc(ProcList)&& ;

        InStr::last =>
                WriStr::last&& ;

        InStr::BadMsg =>
                WriStr::[bad_message,BadMsg].
code

mode find_name(?,?,^,^).

        /*find process named, outputting its input stream and a
          proc list without that process. If not found then fail*/
find_name([proc(Name,ProcInput)|PLshort],Name,ProcInput,PLshort) <-
        true;
find_name([Proc|PLRest],Name,ProcInput,PLshort) <-
        PLshort = [Proc|PLshorter] &
        find_name(PLRest,Name,ProcInput,PLshorter).

mode drawNet(?,^).

drawNet(ProcList,NewProcList) <-
        open(network,write,NetFile)&
        drawProc(ProcList,NetFile,NewProcList)&
        close(NetFile)&
        display_network.

mode drawProc(?,?,^).

drawProc([],_,[]).

drawProc([proc(Name,ProcInput)|Rest],NetFile,
                        [proc(Name,ProcInput1)|Rest1]) <-
        ProcInput = [dump(InstList,ObjLevel)|ProcInput1]&
        writeNet(NetFile,InstList,ObjLevel)&
        drawProc(Rest,NetFile,Rest1).

mode stop_proc(?).

stop_proc([]).

stop_proc([proc(Name,ProcInput)|Rest]) <-
        ProcInput = []&
        stop_proc(Rest).

mode writeMsg(?,?).
%write destination and type of message
```

```
writeMsg(OutFile,msg(NameTo,Msg)) <-
        functor(Msg,MsgType,Arity)&
        writeMy(OutFile,[NameTo,MsgType]).

mode writeCount(?).
%write count if its a multiple of 5
writeCount(Count) <-
        Rem is Count mod 5&
        Rem =:= 0:
        write('''*''')&
        flush_output(user_output);
writeCount(Count).

end.
```

A.2   Give-Way Source Code

In the next example, a *giveWay* concept-frame is shown, that which detects
the existence of a give-way situation between a U-turning car and an oncoming
car.

```
giveWayUt.           %give way to oncoming traffic when U-turning on road
        Out ostream
invisible InstList state <= [], ObjType state <= giveWayUt,
                       ObjLevel state <= 3, OutFile state
initial open(ObjType,write,OutFile)&
        writeMy(OutFile,mmmmmmmmmmmmmmmmmmmmmmmmmmmm)&
        xterm(ObjType)&& .
clauses
        last =>          writeMy(OutFile,[stopping,InstList])&
                close(OutFile)&
                Out::last.      /*dump insts & stop*/

        dump(OutInstList,OutObjLevel)         =>
                OutInstList = InstList&
                OutObjLevel = ObjLevel && ;

        create(Id,Relns)        =>
                getInst(Id,InstList,inst(TargId,Props,Justn),ExcInstList):
                writeList(OutFile,[create,unnecessary,as,Id,found])&
                union(Relns,Props,NewProps)&
                InstList becomes [inst(TargId,NewProps,Justn)|ExcInstList]

                else         /*inst not found */
                writeList(OutFile,[created,Id])&
                checkAssns(Id)& /*send 'check' to likely
                                        associates with this inst Id*/
                InstList becomes
                        [inst(Id,Relns,[])|InstList] && ;
                /*dont process any more msgs until InstList is updated!*/

        check(reln(composedOf,SendId,Bel),Done) =>
                %Note Done flag to delay sender until this object updated
                Done = yes&
                writeList(OutFile,[check,relation,composedOf,from,SendId])&
                getType(SendId,SendType)&
                %get the road the 'right' is in
                Out::msg(SendType,getVal(SendId,
                        reln(composedOf,id(road,any,bel),bel),RoadProps))&
                Out::msg(SendType,getVal(SendId,
```

```
                        desc(from(any),bel),RtFrom))&
                sendRoad(OutFile,InstList,RoadProps,RtFrom,SendId,
                        NewInstList)&

                /*found if FoundIds not empty*/
                InstList becomes NewInstList&& ;

        negCheck(Id) =>
                %Id is removed from the Property lists of all Insts
                writeList(OutFile,[Id,removed,from,all,relations])&
                delRef(InstList,Id,NewInstList)&
                InstList becomes NewInstList&& ;

        anyInst(FoundList,Prop) =>
                writeList(OutFile,[enquiry,regarding,Prop])&
                /*Note: this only checks properties in the PropList*/
                seek(InstList,Prop,FoundList)&& ;

        getVal(WantId,Prop,FoundPropList) =>
                %propagate to other traffic obj
                getInst(WantId,InstList,inst(InstId,Props,Justn),ExcInstList):
                        /*fail if not found*/
                writeList(OutFile,[enquiry,to,WantId,re,Prop])&
                  searchProps(Props,Prop,FoundPropList)&& ;

        getVal(WantId,Prop,FoundPropList) =>
                writeList(OutFile,[enquiry,to,WantId,re,Prop,failed])&
                FoundPropList = []&& ;    /*If Id wrong, return []*/

        updVal(Id,Relns) =>
                getInst(Id,InstList,inst(TargId,Props,Justn),ExcInstList):
                writeList(OutFile,[update,Id,with,Relns])&
                union(Relns,Props,NewProps)&
                InstList becomes [inst(TargId,NewProps,Justn)|ExcInstList]

                else    /*inst not found, so create it but dont checkAssns*/
                writeList(OutFile,[update,of,Id,with,Relns,failed,not,found])&
                InstList becomes
                        [inst(Id,Relns,[])|InstList] && ;

        inq => writeMy(OutFile,[inq,InstList])&& ;

        WrongMsg =>      writeMy(OutFile,[bad_msg,WrongMsg]).
```

code

```
mode checkAssns(?,^,?).
/* check normal associations of this object type to see if they exist in the
   expected relationship.  Note the result of this checking is returned in
   msg 'updVal' */

checkAssns(InstList,InstList,InstId).


/****************************************************************************/
mode sendProps(?,?).
/* send list of properties to another object instance */
sendProps(_,[]);          %if no props, dont send anything
sendProps(id(ToType,ToNum,Bel),Props) <-
        Out::msg(ToType,updVal(id(ToType,ToNum,Bel),Props)).


/****************************************************************************/
mode sendRoad(?,?,?,?,?,^).          %OutFile,InstList,RoadProps,
                                     %RtFrom,SendId,NewIL
%send to straight to get cars from other dirn
sendRoad(OutFile,InstList,[reln(composedOf,id(road,RoadNo,BelI)),Bel)|Rest],
                [desc(from(Dir),BelD)|Rest2],SendId,NewIL) <-
        length(Rest,Len)&
        writeListCond(user_output,Len,[SendId,composed,of,
                more,than,one,inXn,Rest])&
        length(Rest2,Len2)&
        writeListCond(user_output,Len2,[SendId,from,more,
                than,one,dirn,Rest2])&
        Out::msg(straight,anyInst(FoundStr,reln(composedOf,
                id(road,RoadNo,BelI),Bel)))&
        getIdLFromInstL(FoundStr,FoundIds)& %%%ds
        checkTarget(OutFile,InstList,targetOf,FoundIds,
                    SendId,NewIL);

sendRoad(OutFile,InstList,[],_,_,InstList);          %empty composure
sendRoad(OutFile,InstList,_,[],_,InstList);          %empty from list
%error
sendRoad(OutFile,InstList,Wrong1,Wrong2,Wrong3,InstList) <-
        writeList(OutFile,[bad,call,to,sendRoad,in,giveWayUt,ie,Wrong1,
                Wrong2,Wrong3]).


/****************************************************************************/
mode checkTarget(?,?,?,?,?,^). %OutFile,InstList,TargReln,CommStr,SendId,NewIL
%before creating the Inst, check car is pointing toward subject

checkTarget(OutFile,InstList,TargReln,[Id|Rest],SendId,NewIL) <-
        Out::msg(uTurn,getVal(SendId,reln(composedOf,
                id(car,any,bel),bel),SendCarPropList))&
        getType(Id,IdType)&
```

```
        Out::msg(IdType,getVal(Id,reln(composedOf,id(car,any,bel),bel),
                        CarPropList))&
        checkTarget2(OutFile,InstList,TargReln,Id,CarPropList,
                SendId,SendCarPropList,NewIL1)&
        checkTarget(OutFile,NewIL1,TargReln,Rest,SendId,NewIL);

checkTarget(_,InstList,_,[],_,InstList).

/***************************************************************************/
mode checkTarget2(?,?,?,?,?,?,?,^).
%get the single car from the lists CarPropList & RightCarPropList, send
%to Ors to find if car coming targets left, if so call create.
%first check case where sender is target of car to give way to
checkTarget2(OutFile,InstList,targetOf,Id,[reln(composedOf,CarId,Bel)|Rest],
                SendId,[reln(composedOf,CarSendId,Bel2)|Rest1],NewIL) <-
        Out::msg(car,getVal(CarId,desc(reversed,bel),Return))&
        reverseReln(targetOf,Return,RReln)&
        checkOrs(RReln,CarSendId,CarId,FoundProps)&
        condCreateInst(OutFile,InstList,Id,CarId,SendId,CarSendId,
                FoundProps,NewIL);
%next check case where sender is targetting car to give way
checkTarget2(OutFile,InstList,targetting,Id,[reln(composedOf,CarId,Bel)|Rest],
                SendId,[reln(composedOf,CarSendId,Bel2)|Rest1],NewIL) <-
        Out::msg(car,getVal(CarSendId,desc(reversed,bel),Return))&
        reverseReln(targetting,Return,RReln)&
        checkOrs(RReln,CarSendId,CarId,FoundProps)&
        condCreateInst(OutFile,InstList,SendId,CarSendId,
                Id,CarId,FoundProps,NewIL);

checkTarget2(OutFile,InstList,_,Id,[],SendId,_,InstList) <-
        writeList(OutFile,['Error in checkTarget2, got no cars composing',
                Id]);
checkTarget2(OutFile,InstList,-,Id,_,SendId,[],InstList) <-
        writeList(OutFile,['Error in checkTarget2, got no cars composing',
                SendId]).

/***************************************************************************/
mode reverseReln(?,?,^).
%if Return from car contains desc(reversed,bel), ie is non-empty,
        %then put 'B' on the end of input reln and pass back
reverseReln(Reln,[desc(reversed,BelD)],NewReln) <-
        concatStr([Reln,'''B'''],NewReln);
reverseReln(Reln,_,Reln).

/***************************************************************************/
mode condCreateInst(?,?,?,?,?,?,?,^). %OutFile,InstList,SignDesc,
        %StraightId,StrCarId,StDir,SendId,CarSendId,SeDir,FoundProps,NewIL
```

```
%dont create if FoundProps is []
condCreateInst(_,InstList,_,_,_,_,[],InstList);
%test if new components are in existing Inst, create if not
condCreateInst(OutFile,InstList,Id,CarId,SendId,CarSendId,FoundProps,NewIL) <-
        dupeInst(InstList,[reln(composedOf,Id,bel),
        reln(composedOf,SendId,bel)],DupIds)&
        %DupIds is list of Ids of Insts with components
        DupIds =@= []:
        createInst(OutFile,InstList,Id,CarId,SendId,
                CarSendId,[reln2(dum,Id,SendId,bel)],NewIL);  %%ds
%or do nothing
condCreateInst(_,InstList,_,_,_,_,_,InstList).

/****************************************************************************/
mode createInst(?,?,?,?,?,?,?,^).    %OutFile,InstList,SignDesc,StraightId,
                                     %StrCarId,StDir,SendId,CarSendId,
                                     %SeDir,Justn,NewIL
%create new insts for members of CommStr
createInst(OutFile,InstList,Id,CarId,SendId,
                        CarSendId,Justn,NewIL) <-
        getNewOrsAddr(NewNo)&
        belTwoOfN(Justn,NewBel)&
        union([inst(id(giveWayUt,NewNo,NewBel),[reln(composedOf,Id,bel),
                reln(composedOf,CarId,bel),reln(composedOf,SendId,bel),
                reln(composedOf,CarSendId,bel),
                desc(subject(CarId),bel)],Justn)],InstList,NewIL)&
        getVel(CarSendId,Vel)&
        velPhrase(Vel,Phrase)&
        name(NL,[10,13])&
        Out::msg(result,result(['''Give Way to Oncoming: ''',CarSendId,NL,
                ''' U-turning''',Phrase,to,CarId,going,
                straight,''': ''',NL,NewBel]))&
        writeList(OutFile,[adding,inst,NewNo,composed,of,Id,SendId,CarId,
                CarSendId])&
        getNewOrsAddr(NewNo1)&
        Out::msg(giveWay,create(id(giveWay,NewNo1,NewBel),
                [reln(givesWay,CarId,bel),reln(rightOfWay,CarSendId,bel)]))&
        sendProps(Id,[reln(partOf,id(giveWayUt,NewNo,NewBel),bel)])&
        sendProps(SendId,[reln(partOf,id(giveWayUt,NewNo,NewBel),bel)]).

/****************************************************************************/
mode velPhrase(?,^).         %OffendId,velocity phrase
%send msg to result conditional upon speed of OffendId
velPhrase([],''' gives way ''');         %velocity unknown
velPhrase([Vel],'''has given way''') <-
        stopped(Vel,stopped):
        true;
```

```
velPhrase([Vel],'''should have (but hasn't) given way''');
velPhrase(Vel,_) <-
        writeMy([error,in,velPhrase,parameter,is,Vel])&
        fail.

/*************************************************************************/
mode getVel(?,^). %id(Car,CarNo,Bel),[desc(vel(VelX,VelY),BelV)]
%get the velocity desc from car, if not there return []   %%vel
getVel(id(Car,CarNo,Bel3),VelList) <-
        Out::msg(Car,getVal(id(Car,CarNo,Bel3),desc(vel(any,any),bel),
                VelList)).


end.
```

References

[1] AISBETT, J. Optical flow with an intensity-weighted smoothing. *IEEE Transactions on Pattern Analysis and Machine Intelligence 11*, 5 (May 1989).

[2] ALLEN, J. *Natural Language Understanding.* Benjamin/Cummings, Menlo Park, 1987.

[3] ANDERSEN, S. K., OLESEN, K. G., JENSEN, F. V., AND JENSEN, F. HUGIN – a shell for building Bayesian belief universes for expert systems. In *IJCAI-89* (Detroit, August 1989), pp. 1080–1085.

[4] ANDRE, E., HERZOG, G., AND RIST, T. On the simultaneous interpretation of real world image sequences and their natural language descriptions: the system SOCCER. In *ECAI 88. Proceedings of the 8th european conference on artificial intelligence* (UK, 1988), Y. Kodratoff, Ed., pp. 449–54.

[5] BAJCSY, R., JOSHI, A., KROTKOV, E., AND ZWARICO, A. LAND-SCAN: a natural language and computer vision system for analysing aerial images. In *IJCAI 85: Proceedings of the Ninth Intl Joint Conf on Artificial Intelligence* (1985), Int Joint conferences on Artificial Intelligence Inc, Morgan Kaufmann.

[6] BALDWIN, J. Support logic programming. In *Fuzzy sets - Theory and Applications, Proceedings of NATO Advanced Study Institute*, A. Jones et al., Eds. Reidel Pub. Co., Norwell, MA, 1986.

[7] BELL, B., AND PAU, L. F. Context knowledge and search control issues in object-oriented prolog-based image understanding. *Pattern Recognition Letters 13*, 4 (April 1992), 279–290.

[8] BISCHOF, W. F., AND CAELLI, T. Learning structural descriptions of patterns: a new technique for conditional clustering and rule generation. *Pattern Recognition 27*, 5 (1994), 689–697.

[9] BOBICK, A. F., AND BOLLES, R. C. The representation space paradigm of concurrent evolving object descriptions. *IEEE Transactions on Pattern Analysis and Machine Intelligence 14*, 2 (Feb 1992), 146–156.

[10] BODINGTON, R., SULLIVAN, G., AND BAKER, K. Experiments on the use of the ATMS to label features for object recognition. In *Computer Vision - ECCV90* (Antibes, France, April 1990), O. Faugeras, Ed., vol. 427 of *Lecture Notes in Computer Science*, INRIA, Springer Verlag, pp. 542–551.

[11] BOGLER, P. L. Shafer-Dempster reasoning with applications to multisensor target identification systems. *IEEE Transactions on Systems, Man and Cybernetics SMC-17*, 6 (November 1987), 968–977.

[12] BUNKE, H. Hybrid methods in pattern recognition. In *Pattern Recognition Theory and Applications*, P. A. Devijer and J. Kittler, Eds., vol. F30 of *NATO ASI*. Springer-Verlag, 1987.

[13] CAELLI, T., AND DREIER, A. Some new techniques for evidence-based object recognition: EB-ORS1. In *IAPR-92 Proceedings* (Hague, September 1992), pp. 450–455.

[14] CHARNIAK, E., AND MCDERMOTT, D. *Introduction to Artificial Intelligence*. Addison-Wesley, Reading, Massachusetts, 1985.

[15] CIPOLLA, R., AND YAMAMOTO, M. Stereoscopic tracking of bodies in motion. *Image and Vision Computing 8*, 1 (february 1990), 85–90.

[16] CLANCEY, W. J. Situated cognition: How representations are created and and given meaning. In *AERA 1991 Symposium* (Chicago, 1991).

[17] CONLON, T. *Programming in PARLOG*. Addison-Wesley, Menlo Park, Ca., 1989.

[18] DANCE, S., AND CAELLI, T. On the symbolic interpretation of traffic scenes. In *ACCV93 Proceedings of the Asian Conference on Computer Vision* (Osaka Japan, november 1993), pp. 798–801.

[19] DANCE, S., AND CAELLI, T. A symbolic object-oriented picture interpretation network: SOO-PIN. In *Advances in Structural and Syntactic Pattern Recognition, Proceedings of the International Workshop* (Bern, Switzerland, 1993), H. Bunke, Ed., World Scientific Publishing Co., pp. 530–541.

[20] DANCE, S., CAELLI, T., AND LIU, Z.-Q. A network-of-frames model for symbolic scene interpretation. Submitted for publication to Pattern Recognition Journal, August 1994.

[21] DAVISON, A. From Parlog to Polka in two easy steps. In *PLILP'91 : 3rd Int. Symp. on Programming Language Implementation and LP* (Passau, Germany, August 1991), no. 528 in Springer LNCS, Springer, pp. 171–182.

[22] DE KLEER, J. An assumption based TMS. *Artificial Intelligence 28*, 2 (March 1986).

[23] DELLEPIANE, S., SERPICO, S. B., AND VERNAZZA, G. 3D organ recognition by tomographic image analysis. In *Pattern Recognition Theory and Applications*, P. A. Devijer and J. Kittler, Eds., vol. F30 of *NATO ASI*. Springer-Verlag, 1987.

[24] DEMPSTER, A. P. A generalization of Bayesian inference. *Journal of the Royal Statistical Society 30* (1968), 205–247.

[25] DENNETT, D. *Consciousness explained*. Little, Brown and Co., Boston, 1991.

[26] DILLON, C. Image pattern recognition system (IPRS) user manual. Tech. Rep. 93/14, Computer Science, University of Melbourne, Parkville, Victoria, Australia, 1993.

[27] DRAPER, B. A., COLLINS, R. T., BROLIO, J., HANSON, A. R., AND RISEMAN, E. M. The Schema System. *International Journal of Computer Vision 2* (1989), 209–250.

[28] DUBOIS, D., AND PRADE, H. A discussion of uncertainty handling in support logic programming. *International Journal of Intelligent Systems 5*, 1 (March 1990), 15–42.

[29] FELLER, W. *An Introduction to Probability Theory and its Applications*, 3rd ed., vol. 1. Wiley and Sons, New York, 1968.

[30] FERI, R., FORESTI, G., MURINO, V., REGAZZONI, C., AND VERNAZZA, G. Spatial reasoning by knowledge-based integration of visual and IR fuzzy cues. In *Signal Processing V. Theories and Applications. Proceedings of EUSIPCO-90, Fifth European Signal Processing Conference* (1990), L. Torres, E. Masgrau, and M. Lagunas, Eds., Elsevier Amsterdam, Netherlands, pp. 1719–22 vol.3.

[31] GARVEY, T. D. Evidential reasoning for geographic evaluation for helicopter route planning. *IEEE Transactions on Geoscience and Remote Sensing GE-25*, 3 (May 1987), 294–303.

[32] GONZALEZ, R. C., AND WINTZ, P. *Digital Image Processing*. Addison Wesley, Reading, Massachusetts, 1987.

[33] GOVINDARAJU, V., LAM, S. W., NIYOGI, D., SHER, D. B., SRIHARI, R., SRIHARI, S. N., AND WANG, D. Newspaper image understanding. In *Knowledge based computer systems*, S. Ramani, R. Chandrasekar, and K. S. R. Anjaneyalu, Eds. Narosa Publishing House, New Delhi, India, 1990, pp. 375–84.

[34] GREEN, P. E. AF: a framework for real-time distributed cooperative problem solving. In *Distributed Artificial Intelligence*, M. N. Huhns, Ed. Pitman, London, 1987, ch. 6, pp. 153–176.

[35] HAYES, P. J. The naive physics manifesto. In *Expert Systems in the Micro-Electronic Age*, D. Michie, Ed. Edinburgh University Press, Edinburgh, Scotland, 1979.

[36] HERSKOVITS, A. *Language and Spatial Cognition*. Cambridge University Press, 1986.

[37] HEWITT, C. How to use what you know. In *IJCAI-75* (Tbilisi, Georgia, September 1975), pp. 189–198.

[38] HEWITT, C. Viewing control structures as patterns of passing messages. *Artificial Intelligence 8* (1977), 323–363.

[39] HEWITT, C. E. Planner: A language for proving theorems in robots. In *IJCAI'69* (Washington, D.C., 1969).

[40] HOTA, N., RAMSEY, C. L., CHANG, L. W., AND BOOKER, L. B. BaRT manual version 3.0. Tech. Rep. 6778, Naval Research Laboratory, US Navy, Washington DC, Feb 1991.

[41] HUANG, T., KOLLER, D., MALIK, J., OGASAWARA, G., RAO, B., RUSSELL, S., AND WEBER, J. Automatic symbolic traffic scene analysis using belief networks. In *Proc of AAAI-94* (Seattle, August 1994).

[42] KELLER, J., HOBSON, G., WOOTTON, J., NAFARIEH, A., AND LUETKEMEYER, K. Fuzzy confidence measures in midlevel vision. *IEEE Transactions on Systems, Man and Cybernetics SMC-17*, 4 (july 1987), 676–683.

[43] LAKOFF, G. *Women, Fire and Dangerous Things*. University of Chicago Press, 1987.

[44] LOWRANCE, J. D., STRAT, T. M., WESLEY, L. P., GARVEY, T. D., RUSPINI, E. H., AND WILKINS, D. E. The theory, implementation and practice of evidential reasoning. SRI Project 5701, SRI International, Menlo Park, CA 94025, June 1991. Final report.

[45] MATSUYAMA, T., AND HWANG, V. S. *SIGMA A Knowledge based aerial image understanding system.* Plenum Press, New York, 1990.

[46] MCDERMOTT, D. V., AND DOYLE, J. Non-monotonic logic I. *Artificial Intelligence 13*, 1,2 (1980), 41–72.

[47] MINSKY, M. A framework for representing knowledge. In *The psychology of computer vision*, P. H. Winston, Ed. McGraw-Hill, New York, 1975.

[48] MOHNHAUPT, M., AND NEUMANN, B. On the use of motion concepts for top down control in traffic scenes. In *Computer Vision ECCV 90. First european conference on computer vision proceedings* (1990), O. Faugeras, Ed., pp. 598–600.

[49] MULDER, J. A., MACKWORTH, A. K., AND HAVENS, W. S. Knowledge structuring and constraint satisfaction: The MAPSEE approach. Tech. Rep. 87-21, Dept. of computer science, Uni of British Columbia, Vancouver, BC, Canada V6T 1W5, June 1987.

[50] NEUMANN, B. Natural language description of time-varying scenes. In *Semantic Structures: advances in natural language processing*, D. L. Waltz, Ed. Lawrence Erlbaum, Hillsdale, N.J, 1989, pp. 167–207.

[51] NIEMANN, H., BUNKE, H., HOFMANN, I., SAGERER, G., WOLF, F., AND FEISTEL, H. A knowledge based system for analysis of gated blood pool studies. *IEEE Transactions on Pattern Analysis and Machine Intelligence 7*, 3 (may 1985), 246–259.

[52] PEARL, J. *Probabilistic reasoning in intelligent systems: Networks of plausible inference.* Morgan Kaufman, San Mateo, CA, 1988.

[53] PR, C., AND R, K. Information retrieval by constrained spreading activation in semantic networks. *Information Processing and Management 23*, 4 (1987), 255–268.

[54] PROVAN, G. M. An analysis of knowledge representation schemes for high level vision. In *Computer Vision - ECCV90* (Antibes, France, April 1990), O. Faugeras, Ed., vol. 427 of *Lecture Notes in Computer Science*, INRIA, Springer Verlag, pp. 537–541.

[55] REITER, R., AND MACKWORTH, A. A logical framework for depiction and image interpretation. *Artificial Intelligence 41* (1990), 125–155.

[56] RETZ-SCHMIDT, G. Deictic and intrinsic use of spatial prepositions. In *Spatial Reasoning and Multi-Sensor Fusion* (1987), A. Kak and S. Chen, Eds., Morgan Kaufman, pp. 371–380.

[57] RICH, E., AND KNIGHT, K. *Artificial Intelligence*, 2nd ed. McGraw-Hill, New York, 1991.

[58] RINGWOOD, G. A. The dining logicians. Master's thesis, Department of Computing, Imperial College, London, 1984.

[59] ROBINSON, J. A machine-oriented logic based on the resolution principle. *Journal of the Association for Computing Machinery 12*, 1 (1965).

[60] SCHANK, R., AND ABELSON, R. *Scripts, Plans, Goals and Understanding*. Erlbaum, Hillsdale, N.J, 1977.

[61] SCHIRRA, J. R. J., BOSCH, G., SUNG, C. K., AND ZIMMERMANN, G. From image sequences to natural language: a first step toward automatic perception and description of motions. *Applied Artificial Intelligence 1*, 4 (87), 287–305.

[62] SHAPIRO, E., AND TAKEUCHI, A. Object oriented programming in concurrent prolog. In *Concurrent Prolog*, E. Shapiro, Ed., vol. 2. MIT Press, 1987, ch. 29, pp. 251–273.

[63] SLEZAK, P. Situated cognition: Empirical issue, paradigm shift or conceptual confusion. In *Proceedings of the Sixteenth Conference of the Cognitive Sccience Society* (Atlanta, August 1994), A. Ram and K. Eiselt, Eds., Lawrence Erlbaum, Hillsdale, New Jersey, pp. 806–811.

[64] TROPF, AND WALTERS. An ATN for 3D recognition of solids in single images. In *Proceedings of the 8th Int'l Joint Conf. on Artificial Intelligence* (1983).

[65] TSOTSOS, J. K. The complexity of perceptual search tasks. In *Proc. International Joint Conference on Artificial Intelligence* (Detroit, August 1989), N. S. Sridharan, Ed., Morgan Kaufman, pp. 1571 – 1577.

[66] VERRI, A., AND POGGIO, T. Against quantitative optical flow. In *First International conference on Computer Vision* (1987), IEEE, pp. 171–180.

[67] WESLEY, L. P. Evidential knowledge-based computer vision. Tech. Rep. 374, A. I. Centre, SRI International, Menlo Park, CA 94025, January 1986.

[68] WINSTON, P. *Artificial Intelligence*, second ed. Addison-Wesley, Reading, Massachusetts, 1984.

[69] WYLIE JR, C. *Introduction to Projective Geometry.* McGraw-Hill, 1970.

[70] YONEZAWA, A., AND HEWITT, C. Modelling distributed systems. In *IJCAI-77* (Massachusetts, August 1977), Kaufman, pp. 370–376.

Index

A* search, 6
acceleration, 97
accuracy, 106
actors, 17, 24
acyclic networks, 66
affine transformation, 51
Allen, 27
Andre et al, 8
assignment of weights, 68
associative network, 6
assumption-based truth mainte-
        nance system, 7, 84
ATMS, 7, 84
ATN, 5, 6
augmented transition network, 5

background subtraction, 60
Bajcsy et al, 5, 26
Baldwin, 2, 3, 68, 71, 77, 80, 81
BaRT, 67
Bayes, 66, 72
Bayesian belief network, 10
belief, 3, 66
belief and vision, 81
belief interval, 69
belief pair, 69, 78, 81, 82, 84, 106
belief updating, 83

Bell and Pau, 10, 18, 121
belTwoOfN, 77, 79, 83
belUpd, 84
blackboard systems, 18
Bobick and Bolles, 10
Boddington et al, 7, 106
Bogler, 67
bottom-up, 4, 17, 27, 42, 47
Bunke, 28
burglars, 68, 70

car velocities, 52
CARRS, 7
causal support, 66
centroid, 48, 90
Cipolla and Yamamoto, 89
CITYTOUR, 8
Cohen et al, 25
collisions, 96, 100
combining belief, 77
combining evidence, 70, 82
combining independent proposi-
        tions, 71
communication channel, 22
comparing match lists, 92
compatibility mappings, 67
compound objects, 30, 37, 46, 77

concept-frame, 3, 24, 31, 83
concept-instance, 24, 82
conceptual dependency, 17
concurrent execution, 2, 42, 47
concurrent prolog, 20
conditional probabilities, 66
conflicting evidence, 70
conjunction, 72, 76, 79
connected labeled regions, 60
constraint propagation networks, 6
correspondence problem, 3, 89
curvature, 99
cutlery scenario, 3, 35

data channel, 31
data structure, 31, 82
deadlock concept-frame, 53
deletion of instances, 84
Dellepiane et al, 6
Dempster, 70
Dempster-Shafer, 66, 68, 77, 80
Dennett, 119
dependency, 72, 82, 84
disjunction, 74, 76, 79
domain knowledge, 1, 30, 106
Draper et al, 17
Dubois and Prade, 81

evidential reasoning, 67
evidential support, 66
existence criteria, 30, 83

Feller, 77
Feri et al, 9

first order predicate calculus, 20
frame of discernment, 66–68, 71, 78, 81
frames, 3, 16
FRIL, 68
fuzzy confidence measures, 86

Gabor filter, 89
Garvey, 68, 121
geometric scene description, 8
give-way concept-frame, 53
Govindaraju et al, 6
GRANT, 25
graph parsing, 4
Green, 24

Hayes, 30
heading angle, 52
Herskovits, 2, 15, 29
Hewitt, 2, 17, 21, 24, 119
high-level vision, 1, 2
highway traffic scenes, 67
Horn clause, 72, 81
Huang et al, 10, 67
HUGIN, 10, 67
human categorization, 14
hypothesize and test, 2, 4, 6, 10

idealized cognitive models, 14
identity, 82
identity problem, 30
illegality, 94, 100, 106
image sequences, 88
independence, 77, 80

intersection, 51
IPRS library, 60

justification list, 83, 84

Keller et al, 86

Lakoff, 2, 14, 19
linguistic hedges, 86
logic programming, 2, 10, 20, 81
low-level processing, 3, 48, 60
Lowrance et al, 67, 69, 70, 81

Mackworth, 7
major axis, 48, 52, 90
MAPSEE, 7
Markov trees, 66
mass distribution, 67, 69
Matsuyama and Hwang, 9
median filtering, 60
message passing, 31, 84
Minsky, 2, 16, 21, 119
Mohnhaupt and Neumann, 51, 89
Mulder et al, 6
multi-threaded systems, 5
multiple frames, 104

nearness predicate, 29, 85
negative information, 30, 42
negCheck, 84
network-of-frames, 28, 47
Neumann, 8, 18
Nieman et al, 6
non-monotonic reasoning, 30

object orientation, 2, 9, 21
optical flow, 88
orientation, 90, 91

pairing, 90, 91
parlog, 20
parlog++, 3, 23, 28, 31, 78, 81, 93
parsing, 5
Pearl, 66, 80, 81
pedestrian, 97
phase space, 51
PLANNER, 20
plausibility, 69, 76, 82
predicate calculus, 6
probability, 66, 77
procedural subroutines, 84
prolog, 10, 20
Provan, 7
proximity to boundary, 85

real images, 48
red light, 103
Reiter and Mackworth, 7
results, 103
Ringwood, 27
road, 51
Robinson, 20
rotation rates, 92
runtime experiments, 85

scene analysis, 2
scene model, 26
Schank, 2, 17
SCHEMA, 17

Schema, 121
Schirra et al, 8, 86
scripts, 17
segmentation problem, 105
semantic network, 6
SGI Personal Iris, 106
Shapiro and Takeuchi, 2, 22
SIGMA, 9, 18
Sigma, 121
single-threaded systems, 2, 5
situatedness, 29
Slezak, 2, 30
smalltalk, 21
SOCCER, 8
SOO-PIN, 1, 24, 81
sparse image sequences, 89
spatial database, 47, 48
spatial predicates, 29, 66, 84, 93
spatial prepositions, 15
Sun SparcStation II, 106
support, 69, 75, 82
switchboard, 28, 44

T-intersection, 51, 103, 105
targeting, 104
tense, 88, 97
token matching, 2, 88
top-down, 4, 17, 27, 42, 47
traffic, 2, 3, 48, 88, 103
traffic jam, 53
traj concept-frame, 93
trajectory, 90, 92
transition network, 2, 5

Tropf and Walters, 5
Tsotsos, 27
turn activity, 94
turn concept-frames, 51
Tweety, 80

U-turn, 106
uncertainty, 3, 66, 81, 88

velocity, 2, 3, 88, 89, 91, 104, 105
Venn diagram, 77
Verri and Poggio, 88
video images, 3, 88
visual tracking, 89

Wesley, 68
wheels scenario, 3, 42
whyStopped concept-frame, 97, 104
world model, 26
wrong interpretations, 106
wrong side of road, 94

XFIG, 53, 86, 103